BASIC SKILLS IN
MATHEMATICS/Book 1

R W Fox

Edward Arnold

© R W Fox 1974

First published 1974
by Edward Arnold (Publishers) Ltd,
41 Bedford Square,
London WC1B 3DQ

Edward Arnold (Australia) Pty Ltd,
80 Waverley Road, Caulfield East,
Victoria 3145, Australia

Reprinted 1975, 1976, 1977 (twice), 1978, 1979, 1981, 1984

ISBN 0 7131 1849 0

Also available:

Answers to BASIC SKILLS
IN MATHEMATICS/Book 1

By the same author:

Certificate Mathematics

A complete course in 3 books to CSE and O Level.

Mathematical Tables and Data

In collaboration with H A Shaw.

Printed in Great Britain at The Pitman Press, Bath

Contents

1 How many? — Number   **1**
Some facts and figures
Counting by numbers — words and figures — counting exercises
Number — addition and subtraction — missing digits
Magic squares

2 How much? — Money   **17**
Money — addition and subtraction

3 How many? — Algebra   **21**
More counting — collecting terms

4 How many? — Geometry   **23**
Geometry — counting faces, corners, edges — plane figures
Tangrams — playing with shapes
Making designs — counting and drawing

5 How many? — Multiplication   **29**
Number — multiplication up to '4 times'

6 How many? — Algebra   **32**
Substitution — further practice with rules of number

7 How many? — Easy multiplication   **34**
Number and money up to '4 times'
Number — multiplication up to '8 times'

8 How many? — Algebra   **40**
Substitution — 4 rules of number — multiplication up to '8 times'

9 How many? — More easy multiplication   **42**
Number and money up to '8 times'

10 The Circle — More designs   **44**
Parts of the circle — use of compasses — designs

11 How many? — Multiplication   **47**
Number up to '12 times'

12 How many? — Algebra   **51**
Substitution — 4 rules of number — multiplication up to '12 times'

13 How many? — Larger numbers   **53**
Multiplying by 10, 100, 1000 and 20, 30, 40 etc.
Long multiplication — simple multiplication in algebra

14 How many? — Division   **57**
Dividing by 10, 100, 1000 and 20, 30, 40 etc.
Easy division in algebra
Number — short division — long division

15 How long? — Measuring length   **63**
Measuring and counting (dm, cm, mm only)
Measuring exercises

16 Measuring in algebra   **71**

17 **Measuring angles**   72
The right angle — acute and obtuse angles
The protractor — measuring angles
Drawing exercises — geometry things to remember

18 **How much? — Money**   81
Money — multiplication — division

19 **How many? — Decimals**   84
Multiplying and dividing by 10, 100, 1000
Addition and subtraction — multiplication and division

20 **How much? — Weight**   91
Weight — 4 rules using the gram only

Preface

It is the author's opinion that modern approaches to the teaching of mathematics frequently do not pursue a particular skill to the point at which a child reaches the *confidence of knowing* that he has mastered the process. Work assignments are often too brief, sacrificing the acquisition of skill for novelty in the names of 'progress' and 'release from boredom'.

Children do not master essential skills incidentally, as some teachers venture to hope. Lack of ability is often due to lack of *sufficient* experience — perhaps the only motivation really needed is the *opportunity* to learn the skills and, for many, this means repeating the processes often enough.

The extensive exercises in this text (nearly 3500 examples) permit the pupil to dwell on those operations and skills which require more experience before moving on — so avoiding the usual search by the teacher for more material of a similar nature from a variety of textbooks. Some opinions suggest, not without good cause, that mechanical skills have become neglected — this series of books seeks to provide a remedy. There is no shortage of 'activity' material on the market, so the author makes no apologies for the absence of such an approach from this present work. Nevertheless, there is much to keep the pupil's *mind* active.

Book 1 seeks to provide groundwork in counting, using number, calculating in money and also in length and weight, though there is some limitation of the units used in the latter two items. Decimals have been given an early introduction due to modern social demands. These days vulgar fractions are probably less in demand and, because they also cause much difficulty, the author has chosen to accommodate them later in the series (Book 2). Book 1 also provides an introduction into simple algebra and geometry but each is used as a vehicle to further the pupil's experience in basic computation.

The material has been prepared to satisfy the needs of a pupil's mathematical ability rather than his chronological age group. There is ample opportunity for the teacher to practise the professional skills and indeed, such teaching will still be essential with slower pupils. Assistance will almost certainly have to be given to those of poor reading ability.

For those teachers wishing to pursue a 'modern approach' or one of the various 'maths projects', *Basic Skills in Mathematics* will provide a valuable backup course in fundamental processes.

R W F

1 How many? – Number

Some facts and figures

Let us begin by making a list of facts about you, your friends and your school.
Write your answers in the form of sentences as shown in the examples.

1. What is your name?
    Answer: *My name is John Henry Smith.*
2. How old are you?
    Answer: *I am 12 years old.*
3. When were you born?
    Answer: *I was born on 21st of August 19*
4. Where do you live? What is your address?
    Answer: *I live at . . .*
5. What is the name of your school?
6. What is the address of your school?
7. How tall are you? What is your height?
8. How heavy are you? What is your weight?
9. What is the colour of your hair?
10. What is the colour of your eyes?
11. People with light coloured hair may be called *blonde*. People with dark coloured hair may be called *brunette*. People with red coloured hair may be called *auburn*. Which word describes you?
12. How many children are there in your family?
13. How many brothers have you?
14. How many sisters have you?
15. What is the name of your class?
16. What is the number of your class room?
17. What is the name of your class teacher?
18. What is the name of your mathematics teacher?
19. How many children are there in your class?
20. How many boys are there in your class?
21. How many girls are there in your class?
22. How many children are there in your year?
23. How many children are there in your school?
24. How many classes are there in your year?
25. How many classes (or forms) are there in the whole school?

26. How many teachers are there in the school?
27. How many desks are there in your room?
28. How many chairs are there in your room?
29. How many doors are there in your room? Don't forget the cupboards.
30. How many panes of glass are there in your room?
31. How many of the children in your class do you know well?
32. How many question marks are there altogether in these questions?
33. How many of these questions begin with the words *How many*?
34. How many full-stops are there in a bottle of ink?
35. How many beans make five?
36. How long is a piece of string?

Counting by numbers

Probably because we can use our fingers and thumbs in counting, the numbers we use in everyday life are counted by using groups of 10.

> Single items are counted in UNITS up to nine
> 10 units make a TEN
> 10 tens make a HUNDRED
> 10 hundreds make a THOUSAND

NOTE

(i) The names we use are:
UNITS
TENS
HUNDREDS
THOUSANDS
TEN-THOUSANDS
HUNDRED-THOUSANDS
MILLIONS

(ii) In a given **number**, the value of each *figure* depends upon its *position*.

EXAMPLES. *Write the following numbers in words*

	MILLIONS	100 THOUSANDS	10 THOUSANDS	THOUSANDS	HUNDREDS	TENS	UNITS
A				4	3	2	1
B				5	0	0	2
C			5	6	7	8	9
D		1	0	3	4	6	0
E	1	0	2	0	3	0	4
F	5	0	0	0	0	0	2

Answers
(A) four thousand, three hundred and twenty-one
(B) five thousand and two
(C) fifty-six thousand, seven hundred and eighty-nine
(D) one hundred and three thousand, four hundred and sixty
(E) one million, twenty thousand, three hundred and four
(F) five million and two

NOTE When writing large numbers, small spaces are left to show multiples (or groups) of a thousand, e.g. 103 460; 1 020 304; 5 000 002

Exercise 1

Write the following numbers in words:

1. 9	**2.** 7	**3.** 8	**4.** 3
5. 12	**6.** 18	**7.** 14	**8.** 11
9. 15	**10.** 13	**11.** 19	**12.** 16
13. 20	**14.** 17	**15.** 25	**16.** 29
17. 23	**18.** 26	**19.** 21	**20.** 24
21. 27	**22.** 22	**23.** 28	**24.** 30
25. 38	**26.** 45	**27.** 32	**28.** 49
29. 54	**30.** 37	**31.** 43	**32.** 51
33. 36	**34.** 60	**35.** 71	**36.** 83
37. 95	**38.** 72	**39.** 64	**40.** 98

41.	186	42.	267	43.	579	44.	401
45.	341	46.	325	47.	607	48.	907
49.	5018	50.	3400	51.	7352	52.	6004
53.	8031	54.	2905	55.	13 303	56.	24 004
57.	57 981	58.	101 101	59.	324 007	60.	3 405 016

Exercise 2

Write the following numbers in figures:

1.	nineteen	2.	seventeen
3.	eight	4.	twenty-five
5.	fifteen	6.	four
7.	twenty-three	8.	thirteen
9.	fourteen	10.	forty-two
11.	thirty-seven	12.	five
13.	fifty-eight	14.	twenty-four
15.	eighteen	16.	thirty-nine
17.	forty-three	18.	fifty-four
19.	thirty-eight	20.	three
21.	twelve	22.	twenty-eight
23.	eleven	24.	forty-seven
25.	sixteen	26.	thirty-three
27.	twenty-one	28.	fifty
29.	forty-eight	30.	twenty-seven
31.	twenty	32.	fifty-two
33.	forty-nine	34.	thirty
35.	fifty-six	36.	thirty-five
37.	nine	38.	fifty-three
39.	twenty-two	40.	forty-four
41.	thirty-four	42.	sixty
43.	forty-five	44.	fifty-one
45.	twenty-six	46.	fifty-five
47.	thirty-two	48.	forty
49.	one	50.	fifty-nine
51.	forty-one	52.	ten
53.	seven	54.	twenty-nine
55.	forty-six	56.	two
57.	fifty-seven	58.	thirty-one
59.	six	60.	thirty-six
61.	two hundred and seventy-three	62.	six hundred and eighty-nine
63.	one hundred and nine	64.	eight hundred and sixty
65.	five hundred and seventy-four	66.	nine hundred and ninety-nine
67.	three hundred and seventy-five	68.	seven hundred and eight

69. four hundred	**70.** one-thousand, eight hundred and thirty-one
71. three thousand and fourteen	**72.** nine thousand and nine
73. ten thousand, five hundred and five	**74.** twenty-five thousand, two hundred and fifty
75. fifty-five thousand and fifteen	**76.** six hundred and fifty-four thousand, three hundred and twenty-one
77. two hundred and five thousand, two hundred and five	**78.** seven hundred thousand and seventy
79. nine hundred and ninety thousand and nine	**80.** fifty-six million, three hundred and forty thousand and ninety-nine

Exercise 3

In each of the following, say whether the statement is *TRUE* or *FALSE*.

1. One thousand and one = 1001
2. Eleven thousand and one = 1101
3. 1101 = eleven hundred and one
4. 9011 = nine thousand and eleven
5. 21 001 = two thousand, one hundred and one
6. Four hundred and four thousand and four = 400 404
7. Fifty-seven thousand, eight hundred = 578 000
8. One hundred and one thousand and ten = 101 010
9. 320 500 = three hundred and twenty thousand, five hundred
10. 11 001 100 = eleven hundred thousand, eleven hundred.

Exercise 4

State the **value** of each of the figures which have been *underlined* in the following:

1. 2<u>4</u>	**2.** <u>5</u>6	**3.** 1<u>8</u>	**4.** 1<u>2</u>3
5. 8<u>9</u>	**6.** <u>3</u>26	**7.** <u>8</u>7	**8.** 2<u>7</u>8
9. 46<u>2</u>	**10.** <u>5</u>13	**11.** 1<u>3</u>23	**12.** 2561
13. 8<u>1</u>32	**14.** 37<u>1</u>3	**15.** <u>6</u>414	**16.** 391<u>7</u>
17. 17<u>9</u>1	**18.** 2<u>2</u>56	**19.** 46<u>2</u>8	**20.** 7<u>1</u>01
21. 4<u>3</u>06	**22.** 43 0<u>6</u>0	**23.** 46 <u>0</u>30	**24.** 36 00<u>4</u>
25. 30<u>6</u> 040	**26.** 30 6<u>0</u>4	**27.** 340 00<u>6</u>	**28.** 6<u>3</u>4 000
29. 3 00<u>6</u> 040	**30.** <u>4</u> 360 000	**31.** 6 <u>4</u>00 300	**32.** <u>4</u>3 060 000

Counting

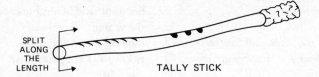

SPLIT ALONG THE LENGTH

TALLY STICK

The first form of shop-keeping began when people 'swopped' things of equal value to each other. To keep a check on the number of items exchanged, traders used a tally-stick.

This was a length of wood with notches scored across it as a tally or check of the goods, narrow cuts for units — wider and deeper ones for tens, and larger cuts still for hundreds. At the end of the trading, the stick would be split down the middle along its whole length, each of the two traders keeping one of the halves.

NOTE You will use this paragraph in question 2, Exercise 5.

Even today a cricket umpire may keep a tally of the number of times a bowler delivers the ball by counting pebbles or coins in the pocket of his coat.

NOTE You will use this paragraph in question 3.

A useful method of keeping a tally is shown on the opposite page. The tally marks should be arranged so that the fifth stroke is placed across the previous four strokes to form a 'gate', each gate is then worth five, thus ⊔⊔⊤.

NOTE Digit is another word meaning figure.
    There are ten digits: 0, 1, 2, 3, 4, 5, 6, 7, 8, 9

EXAMPLE Count the number of times each of the *digits* is used in Exercise 1, including the question numbers.
    The digits of the first ten questions have been tallied for you; copy the table and the title and complete the tally for all the digits (figures) used in Exercise 1.

Digit	Tallies	Total
0	/	
1	~~//// ~~ / / / /	
2	/ /	
3	/ / /	
4	/ /	
5	/ /	
6	/	
7	/ /	
8	/ / /	
9	/ /	

TALLY TABLE TO SHOW THE NUMBER OF TIMES EACH
DIGIT IS USED IN EXERCISE 1.

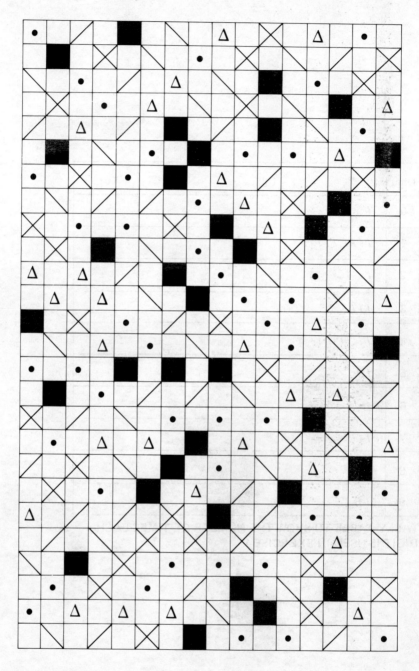

More tallies for you to do

Exercise 5

1. Alternate squares (every other one) of the chart are occupied by one of the following symbols: ⊡ ■ ☑ ◻ ⊠ ◪. How many squares are empty?

   Make up a tally-table to help you count the number of times each symbol occurs. Tally by moving across the rows.

2. Using the paragraph 'This was a length of wood . . . one of the halves' (page 6), make up a tally table to help you count the number of times the following letters occur in the paragraph: a, e, i, o, u, t.

3. Using the paragraph 'Even today . . . his coat' (page 6), make up a tally table to help you count the number of times the following letters occur in the paragraph: f, g, h, m, x, y.

Number — addition and subtraction

Exercise 6

Examine each of the following groups of numbers and write down the next TWO numbers in each case:

1.	1, 2, 3, 4, . . ., . . .,	**2.**	2, 4, 6, 8, . . ., . . .,
3.	3, 5, 7, 9, . . ., . . .,	**4.**	1, 4, 7, 10, . . ., . . .,
5.	5, 8, 11, 14, . . ., . . .,	**6.**	7, 11, 15, 19, . . ., . . .,
7.	23, 22, 21, 20, . . ., . . .,	**8.**	19, 17, 15, 13, . . ., . . .,
9.	41, 43, 45, 47, . . ., . . .,	**10.**	32, 34, 36, 38, . . ., . . .,
11.	16, 19, 22, 25, . . ., . . .,	**12.**	17, 20, 23, 26, . . ., . . .,
13.	39, 36, 33, 30, . . ., . . .,	**14.**	51, 48, 45, 42, . . ., . . .,
15.	21, 25, 29, 33, . . ., . . .,	**16.**	52, 56, 60, 64, . . ., . . .,
17.	1, 6, 11, 16, . . ., . . .,	**18.**	30, 26, 22, 18, . . ., . . .,
19.	42, 37, 32, 27, . . ., . . .,	**20.**	111, 114, 117, 120, . . ., . . .,
21.	18, 14, 10, . . ., . . .,	**22.**	33, 29, 25, . . ., . . .,
23.	10, 20, 30, . . ., . . .,	**24.**	30, 50, 70, . . ., . . .,
25.	100, 90, 80, . . ., . . .,	**26.**	5, 11, 17, . . ., . . .,
27.	26, 32, 38, . . ., . . .,	**28.**	8, 16, 24, . . ., . . .,
29.	9, 17, 25, . . ., . . .,	**30.**	3, 10, 17, . . ., . . .,
31.	14, 21, 28, . . ., . . .,	**32.**	18, 24, 30, . . ., . . .,
33.	22, 33, 44, . . ., . . .,	**34.**	36, 48, 60, . . ., . . .,
35.	56, 48, 40, . . ., . . .,	**36.**	15, 23, 31, . . ., . . .,
37.	32, 25, 18, . . ., . . .,	**38.**	53, 43, 33, . . ., . . .,
39.	0, 1, 3, 6, 10, . . ., . . .,	**40.**	0, 2, 6, 12, 20, . . ., . . .,
41.	29, 22, 16, 11, 7, . . ., . . .,	**42.**	0, 2, 6, 14, 30, . . ., . . .,

Exercise 7

Add the following: (Check your work by adding up and down)

1. 16	**2.** 13	**3.** 22	**4.** 20
10	11	13	15
12	24	14	32
—	—	—	—
5. 32	**6.** 23	**7.** 31	**8.** 25
15	34	41	33
21	40	17	11
—	—	—	—
9. 13	**10.** 21	**11.** 32	**12.** 11
14	24	24	23
15	25	16	37
—	—	—	—
13. 21	**14.** 34	**15.** 25	**16.** 17
36	42	33	33
23	14	22	40
—	—	—	—
17. 23	**18.** 17	**19.** 45	**20.** 38
34	33	33	41
43	50	22	21
—	—	—	—
21. 27	**22.** 31	**23.** 25	**24.** 14
33	45	35	36
42	28	45	58
—	—	—	—
25. 22	**26.** 11	**27.** 9	**28.** 37
34	33	20	14
46	57	72	50
—	—	—	—
29. 19	**30.** 26	**31.** 28	**32.** 39
38	48	38	49
47	27	36	18
—	—	—	—
33. 23	**34.** 32	**35.** 24	**36.** 28
45	38	36	42
67	41	67	65
—	—	—	—

37. 223	38. 135	39. 314	40. 602
322	233	231	135
454	321	413	242
——	——	——	——

41. 123	42. 234˙	43. 237	44. 515
245	152	332	123
332	214	431	362
——	——	——	——

45. 123	46. 245	47. 531	48. 192
436	366	137	245
329	314	248	326
——	——	——	——

49. 234	50. 179	51. 338	52. 769
321	436	553	876
538	567	745	968
——	——	——	——

Exercise 8

The following are **subtractions** or 'take-aways'; you can check your answers by adding the two lower lines, the result should be the same as the top line.

1. 23	2. 34	3. 26	4. 37
11	22	14	25
—	—	—	—

5. 37	6. 46	7. 35	8. 47
16	23	24	25
—	—	—	—

9. 32	10. 41	11. 37	12. 55
10	21	34	45
—	—	—	—

13. 27	14. 35	15. 43	16. 39
24	34	43	24
—	—	—	—

17. 38	18. 37	19. 48	20. 59
15	22	37	45
—	—	—	—

21. 49	22. 38	23. 57	24. 49
36	24	32	21
—	—	—	—

25. 58	26. 57	27. 59	28. 69
32	20	31	22

29. 77	30. 89	31. 96	32. 99
12	31	10	23

33. 124	34. 247	35. 473	36. 695
103	132	121	433

37. 23	38. 34	39. 62	40. 84
14	25	53	75

41. 23	42. 34	43. 62	44. 84
19	28	45	66

45. 57	46. 95	47. 76	48. 48
39	79	57	29

49. 43	50. 32	51. 51	52. 35
26	18	14	16

53. 64	54. 76	55. 87	56. 93
27	28	48	65

57. 132	58. 244	59. 284	60. 347
17	125	168	218

61. 384	62. 827	63. 433	64. 552
195	238	224	349

65. 685	66. 701	67. 852	68. 936
496	397	485	589

69. 375	70. 451	71. 564	72. 733
296	374	487	648

73. 542	74. 631	75. 707	76. 818
539	622	699	799

Exercise 9

The following are **additions**, supply the missing digits (figures):

1.	2.	3.	4.
*1	*2	*3	*4
2*	3*	2*	3*
34	43	45	55

5.	6.	7.	8.
3*	4*	5*	6*
*2	*3	*1	*4
65	67	85	99

9.	10.	11.	12.
*3	*5	4*	5*
41	32	35	22
6*	7*	*8	*9

13.	14.	15.	16.
5*	*3	43	38
*7	49	2*	5*
81	7*	*2	*5

17.	18.	19.	20.
24	4*	*2	87
*6	27	39	4*
4*	*3	66	52
102	125	15*	1*7

21.	22.	23.	24.
13*	251	34*	762
2*7	*63	549	*9*
329	1*6	618	6*9
*11	91*	1**7	*335

Exercise 10

The following are **subtractions** (take-aways), supply the missing digits:

1.	2.	3.	4.
3*	*6	*9	*3
*3	4*	63	5*
22	34	2*	12

5.	6.	7.	8.
4*	5*	6*	68
*6	*2	*5	3*
23	37	14	*3

9.	10.	11.	12.
*4	*3	5*	*6
7*	64	*5	37
15	2*	16	3*

13.	14.	15.	16.
6*	7*	*4	*7
*8	*2	4*	6*
23	39	47	29

17.	235	18.	4*3	19.	*3*	20.	7*6
	9		29*		479		*6*
	*9		*82		*5		58

21.	37*	22.	5*4	23.	*36	24.	34*
	1*3		*39		2*7		*23
	*08		22*		14*		*4

Magic Squares

In each of the following sets of numbers, add the digits:

(i) **Horizontally** (in rows ⟷)
(ii) **Vertically** (in columns ↕)
(iii) **Diagonally** (from corner to corner)

4	3	8
9	5	1
2	7	6

A

7	6	11
12	8	4
5	10	9

B

8	6	16
18	10	2
4	14	12

C

Questions

1. How many answers are there for each square?
2. What do you notice about the answers for each square?
3. Copy the squares and complete the following: *Answer* A =   *Answer* B =
   Answer C =

These are examples of **Magic Squares** and A is called a 'basic square' because others can be formed from it. Compare B with A, the new digits have been obtained by adding 3 to each of the original digits. Compare C with A and you should be able to see how the new digits have been formed. Further squares can be developed by changing the position of the digits in A. (See example D.)

2	9	4
7	5	3
6	1	8

D

4	9	5	16
15	6	10	3
14	7	11	2
1	12	8	13

E

15	16	22	3	*
8	*	20	21	2
*	7	13	*	25
24	*	6	*	18
*	23	4	10	11

F

Exercise 11

1. A and D are only two of the eight possible ways of arranging the same digits to form magic squares. Find two other arrangements for the digits and check the results by adding the rows, columns and diagonals.
2. E is a magic square containing 16 'cells'. Find two other arrangements for the digits and check that the results give magic squares.
3. F is a magic square containing 25 'cells', supply the missing digits.
4. Which of the following are magic squares?

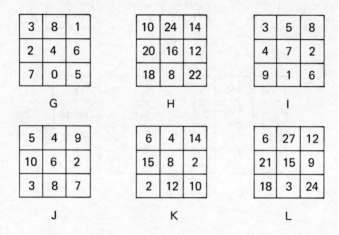

3	8	1
2	4	6
7	0	5

G

10	24	14
20	16	12
18	8	22

H

3	5	8
4	7	2
9	1	6

I

5	4	9
10	6	2
3	8	7

J

6	4	14
15	8	2
2	12	10

K

6	27	12
21	15	9
18	3	24

L

5. Supply the missing digits in the following magic squares:

3	*	7
8	4	*
*	6	5

M

*	2	7
8	*	0
*	6	5

N

6	4	*
16	8	*
2	12	*

O

*	12	7
10	8	*
9	*	11

P

4	18	8
*	*	*
12	2	16

Q

10	8	18
20	*	4
*	16	*

R

6	7	*
*	5	9
8	*	4

S

8	6	*
*	10	2
4	*	12

T

*	2	7
4	*	8
*	10	3

U

12	*	24
27	15	*
*	21	18

V

15	18	*
*	12	24
21	*	9

W

28	*	20
8	16	*
12	*	4

X

3	*	4	15
*	5	9	2
13	6	10	*
0	11	*	12

Y

*	26	28	6
*	*	10	16
*	20	18	8
*	2	4	*

Z

2 How much? – Money

Addition and Subtraction

NOTE (i) There are 100 pence (**p**) in 1 pound (**£**).
  (ii) A dot (**decimal point**) is used to separate the new pence from the
     pounds (£3.25). When printed this is usually put above the line (£3·25)
     but, when typed, it is put on the line as in this book (£3.25). In handwriting
     a dash is often used.
  (iii) In money sums the digits (figures) are added or taken away just as
     they are with ordinary numbers but we must not forget the decimal
     points. **The points must be kept under each other to separate the pence
     from the pounds (£s).**

Exercise 12

Add the following:

£		£		£		£
1. 2.31		2. 4.24		3. 3.18		4. 4.19
1.42		2.43		4.52		3.32
3.15		1.58		2.37		2.53
£		£		£		£

£		£		£		£
5. 1.52		6. 3.47		7. 4.55		8. 5.50
3.34		2.59		3.45		3.05
4.57		1.91		3.51		2.18
£		£		£		£

£		£		£		£
9. 4.82		10. 3.12		11. 5.72		12. 7.35
2.15		4.58		1.36		2.42
5.37		6.16		2.45		3.18
£		£		£		£

	£		£		£		£
13.	4.35	14.	3.57	15.	6.22	16.	7.55
	8.42		5.31		3.75		4.06
	2.49		6.28		4.93		3.10
	£		£		£		£

	£		£		£		£
17.	8.54	18.	6.34	19.	5.50	20.	8.22
	3.05		7.07		6.04		5.05
	4.08		8.08		7.30		6.40
	£		£		£		£

	£		£		£		£
21.	0.95	22.	0.44	23.	0.32	24.	0.82
	0.32		0.55		0.43		0.26
	0.47		0.66		0.95		0.55
	£		£		£		£

	£		£		£		£
25.	0.42	26.	0.65	27.	0.54	28.	0.47
	0.31		0.72		0.82		0.39
	0.58		0.29		0.19		0.25
	£		£		£		£

	£		£		£		£
29.	0.05	30.	1.20	31.	4.05	32.	5.50
	1.09		3.08		0.06		0.05
	3.00		5.02		3.60		0.80
	£		£		£		£

	£		£		£		£
33.	5.08	34.	6.60	35.	7.35	36.	9.71
	8.24		9.08		8.53		0.09
	7.30		0.92		9.07		7.60
	4.00		4.45		2.80		8.00
	£		£		£		£

	£		£		£		£
37.	2.50	38.	4.08	39.	5.60	40.	0.82
	8.45		6.82		7.27		9.15
	0.98		8.30		6.82		5.00
	6.04		0.86		0.08		8.08
	£		£		£		£

Exercise 13

The following are **subtractions** (take-aways):

£	£	£	£
1. 3.48	2. 4.56	3. 5.82	4. 6.46
2.25	3.34	2.61	4.34
£	£	£	£

£	£	£	£
5. 8.64	6. 9.75	7. 7.83	8. 5.97
4.22	7.53	6.43	3.84
£	£	£	£

£	£	£	£
9. 4.34	10. 5.41	11. 6.28	12. 8.63
2.29	3.25	4.09	6.48
£	£	£	£

£	£	£	£
13. 7.91	14. 6.54	15. 9.28	16. 4.35
5.79	3.28	7.28	2.16
£	£	£	£

£	£	£	£
17. 3.27	18. 3.27	19. 3.27	20. 3.27
2.17	2.19	2.28	0.27
£	£	£	£

£	£	£	£
21. 4.13	22. 5.06	23. 6.21	24. 7.37
2.09	3.18	4.34	5.28
£	£	£	£

£	£	£	£
25. 8.24	26. 9.46	27. 3.72	28. 4.38
7.15	8.47	2.68	3.39
£	£	£	£

£	£	£	£
29. 5.24	30. 6.46	31. 7.72	32. 8.38
4.35	5.59	6.84	8.29
£	£	£	£

19

	£		£		£		£
33.	1.00	**34.**	1.60	**35.**	2.00	**36.**	2.50
	0.22		0.72		0.88		0.94
	£		£		£		£

	£		£		£		£
37.	3.00	**38.**	4.00	**39.**	5.00	**40.**	5.00
	0.08		0.64		0.10		0.06
	£		£		£		£

Exercise 14

These are mixed 'adds' and 'take-aways'. First add up the *'adds'* then add up the *'take-aways'*. Take the second answer from the first.

EXAMPLE   36p + £2.60 − £1.45 + 85p − £2.16 + £1.72 − £1.80

	£		£		£		
Add	0.36	*Add*	1.45	Take the	5.53		
the	2.60	*the*	2.16	second answer	5.41		
'adds'	0.85	*'take-aways'*	1.80	from the first =	£0.12	or	12p
	1.72		£5.41				
	£5.53						

Now do these:

1. 28p + £1.62 − 85p + £2.54 − £0.65 − £1.47
2. £4.84 − £2.36 + £1.60 − 96p + 82p − £2.08
3. £0.36 + £3.05 − £1.84 − £1.32 + 48p − 52p
4. 78p + £5.00 − £2.20 + £1.08 − £1.85 − £2.32
5. £3.00 − £1.50 + £2.38 − £2.30 + 85p − 95p
6. 92p + £3.25 + £4.64 − £1.62 − £2.16 − 78p − £1.06
7. £4.21 − £1.18 − £3.24 − £1.08 − £1.15 + £5.00
8. 85p + 68p + 73p + 16p + 45p + 37p − £2.18
9. £5.00 − 16p − 27p − 38p − 49p − 62p − 51p
10. 37p + 16p + 21p + 48p + 54p − 22p − 10p − 61p
11. £120 − £60 + £34 + £28 − £47 − £52 − £36 + £40
12. £10.07 + £11.05 − £9.20 − £3.00 − 92p + £5.40
13. £11.64 − £15.23 − £23.62 − £18.24 + 68p + £50
14. £22.04 + £13.06 + £5.09 + £16.20 + £3.01 − £21.50
15. £10.20 + £4.00 + £0.47 − 98p − 95p − £3.38
16. £15.00 − £2.08 − £3.20 − £4.30 − £5.42

3 How many? – Algebra

In **algebra**, we use letters as well as figures, just as we did in Exercise 14.

EXAMPLES

(i) $2a + 3a + 4a + 5a + 6a = 20a$
(ii) $16x - 12x = 4x$
(iii) $4a + 6a - 2a - 8a + 3a = 13a - 10a = 3a$

NOTE (i) When a letter is written on its own it means **one** of them: a means **1**a
                                                                      x means **1**x

(ii) It will help you if you can work out the sums horizontally ($\longleftrightarrow$) as shown in the examples.

Exercise 15

1. $a + a$
2. $a + a + a$
3. $a + a + a + a$
4. $b + b$
5. $b + 2b$
6. $2b + 2b$
7. $b + b + 2b$
8. $a + 2a + a$
9. $2a + a + a$
10. $c + c + c$
11. $2c + 2c + 2c$
12. $c + 2c + 3c$
13. $3x + 2x + x$
14. $x + 4x + x$
15. $2x + 3x + 4x$
16. $2y + 4y + 6y$
17. $y + 3y + 5y$
18. $3y + 6y + 9y$
19. $4z + 8z + 12z$
20. $3z + 5z + 7z$
21. $10z + 20z + 30z$
22. $2a + 5a + 9a$
23. $b + 6b + 9b$
24. $6c + 8c + c$
25. $10a - 5a$
26. $12a - 6a$
27. $8a - 5a$
28. $3a - 2a$
29. $5b - 4b$
30. $8b - 7b$
31. $10c - 9c$
32. $7x - 6x$
33. $2y - y$
34. $4g - g$
35. $5x - 3x$
36. $7a - 4a$
37. $8b - 3b$
38. $6p - 5p$
39. $3y - 3y$
40. $16a - 9a$
41. $23b - 13b$
42. $19c - 5c$
43. $5x - 5x$
44. $12y - 4y$
45. $20z - 16z$
46. $24p - 18p$
47. $18q - 9q$
48. $30r - 20r$
49. $2a + a - a$
50. $3b + 2b - b$
51. $2x + 3x - x$
52. $2c + 2c - 2c$
53. $3p + 3p - 3p$
54. $4q + 4q - 4q$
55. $2f + 3f - 2f$
56. $3g + 4g - 3g$
57. $2g + 4g - 2g$
58. $4x - 3x + 2x$
59. $5y - 2y + 3y$
60. $6b - 4b + 2b$

More tallies for you to do

Sometimes a piece of algebra may have several letters in it. When this happens the same letters are collected together by adding or taking away, (+) or (−). We deal with the letters in alphabetical order.

EXAMPLES (i) $3a + 4b + 2a - a + 3b - 2b = 4a + 5b$
(ii) $4x + 3y - 2z - x + 2y + 5z = 3x + 5y + 3z$

NOTE (i) The letters used are called **terms**.

(ii) This kind of tallying is called **collecting like terms**.

Exercise 16

Collect like terms:

1. $a + a + a + b + b + b$
2. $a + b + a + b + a + b$
3. $a + a + b + b + a + b$
4. $b + b + b + a + a + a$
5. $a + b + b + b + b + a$
6. $b + a + a + a + a + b$
7. $b + b + c + c + c + c$
8. $c + b + b + c + b + b$
9. $x + 2x + 3x + y + 2y$
10. $x + 4x + y + 2y + 3y$
11. $2y + 3y + 2x + 3x + x + y$
12. $2x + 4y + 3x + 5y + 4x + 6y$
13. $a + a + b + b + c + c$
14. $a + b + c + a + b + c$
15. $a + 2b + 3c + a + 2b + 3c$
16. $4c + 3c + 2b + b + 3a + 2a$
17. $2a + b + c + c + b + 2a$
18. $3c + 2a + b + 2b + a + c$
19. $2r + 3p + 4p + 3r + p + r$
20. $4f + g + f + 2h + 3g + h$
21. $3h + 2f + 3f + 4f + f + 2g$
22. $a + 2c + 5b + 2a + 3c + b$
23. $3a + b + b - 2a + 2b - b$
24. $2a + 3b - 2a - 2b + a + b$
25. $4a + b - a - b - a + 2b$
26. $2x + x + y + 3x + 2y - 3y$
27. $a + b + 2a + 2b + b$
28. $3a + 3b + 3c - 2a - 2b - 2c$
29. $4c + 4b + 4a - 2c - 2b - 2a$
30. $5a - 3a + 5b - 3b + 5c - 3c$
31. $6a + 6b + 6c - 4c - 4b + 4c$
32. $7a - 5c - 5b - 5a + 7c + 7b$
33. $5x + 3y + z + z - 3y - 5x$
34. $4x - 4x + 4y - 4y + 4z - 4z$
35. $6x - 2y - 4z + 4y + 6z - x$
36. $8z - 5x + 3y + 7x - y - 6z$

4 How many? – Geometry

Here are some simple, solid shapes. In each case they have faces (**surfaces**), in most cases they have edges and corners; the edges which are hidden from you are shown by broken lines.

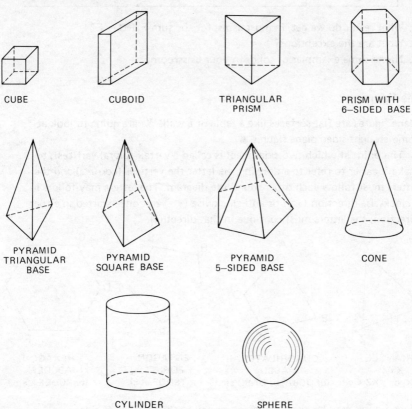

| CUBE | CUBOID | TRIANGULAR PRISM | PRISM WITH 6–SIDED BASE |

| PYRAMID TRIANGULAR BASE | PYRAMID SQUARE BASE | PYRAMID 5–SIDED BASE | CONE |

| CYLINDER | SPHERE |

Exercise 17

For each of the solids shown:

1. Count the number of faces; in some cases they will be flat surfaces, in other cases they will be curved.

2. Count the number of corners.
3. Count the number of edges.
4. Make a table of the information, as shown below and carry out the sum given in the fifth column. F stands for the number of **faces** (surfaces), C stands for the number of **corners** and E stands for the number of **edges**.

Name of solid figure	No. of faces surfaces F	No. of corners C	No. of edges E	Value of F + C − E
Cube	6	8	12	6 + 8 − 12 = 2
Cuboid	6	8	12	
Copy the rest				

5. What result do we get, in most cases, for the sum F + C − E?
6. What are the exceptions?
7. Name some examples of solids in your classroom.

Plane figures

Plane figures are flat surfaces like a table or a wall. We are going to look at some straight-sided plane figures.

The point at which two sides meet is called a **vertex** (plural **vertices**); to make it easier to refer to a drawing, we letter the vertices (corners) and the letters must follow each other round the diagram. The letters may follow in a clockwise direction (⌢) or anti-clockwise (⌢) but once started in either direction, the letters must continue in that direction.

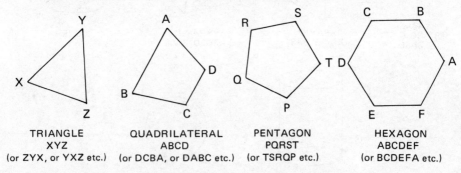

TRIANGLE	QUADRILATERAL	PENTAGON	HEXAGON
XYZ	ABCD	PQRST	ABCDEF
(or ZYX, or YXZ etc.)	(or DCBA, or DABC etc.)	(or TSRQP etc.)	(or BCDEFA etc.)

Plane figures may be given names which tell us the number of sides:—

tri means	3	quad means	4
penta means	5	hexa means	6
hepta means	7	octa means	8
nona means	9	deca means	10

The general name for a 'many-sided' figure is a **polygon**.

Exercise 18

1. Make up a table of information (see below) about each of the plane figures we have mentioned. Letter the vertices (corners).

DRAWING OF PLANE FIGURE	NAME	NO. OF SIDES	NO. OF VERTICES
	Triangle	3	3

2. What have you discovered about the number of sides and the number of corners (vertices) of a plane figure?
3. How many sides has a bee's cell?
4. How many sides has the face of a 50p coin?
5. In the Olympic Games there is an event called the Pentathlon. How many different skills are required for this event?
6. How many skills are required in the Decathlon event?
7. Write down some more words beginning with tri- (meaning 3).
8. What is a biped? Write down another word with a similar meaning.
9. What is a quadruped? Write down another word with a similar meaning.
10. Name four plane surfaces in your classroom.

More plane figures

The diagram on page (26) consists of many lines; separate figures can be identified by using the letters at the vertices of the figure you have in mind. For example:—

RYP is a triangle; AEQL is a quadrilateral. Note that, with so many letters on the diagram, the letters for a particular figure may not follow each other in alphabetical order but you must read them as they appear round the figure.

In the following exercise, name the figures required by using the letters at the vertices (*corners*).

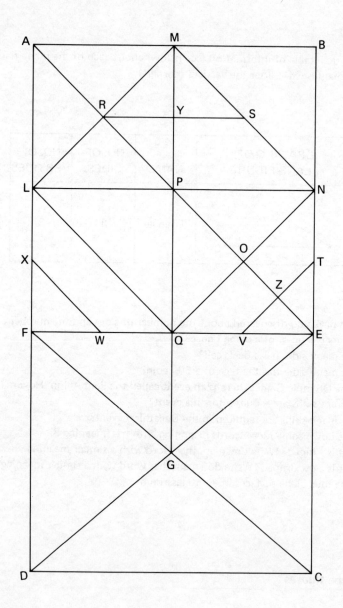

Exercise 19

From the diagram above
1. Name four triangles the same size and shape as triangle RYP.
2. Name three more triangles, all different in size.
3. Name another quadrilateral the same size and shape as AEQL.
4. Name three more quadrilaterals the same shape as AEQL but different in size from each other and from AEQL.

5. Name another quadrilateral the same size and shape as APQL.
6. Name another triangle the same size and shape as DGC.
7. What name can be given to the figure PNTVQ?
8. Name another figure the same shape and size as PNTVQ.
9. Name two more figures the same shape as PNTVQ but each larger in size.
10. Name two hexagons.
11. Name one heptagon.
12. Name one octagon.
13. How many triangles can you count on the diagram?
14. Although we haven't discussed the **square**, which is a special type of quadrilateral, you probably know what a square is when you see one. How many squares can you count on the diagram?

Tangrams — Playing with shapes

Tangrams is a traditional Chinese pastime. A square of card is marked and cut into the shapes shown in the diagram. By rearranging the pieces, many different figures may be made.

If a square is drawn on squared paper, as shown, the marking out will be made easier. The paper may then be stuck to card before the shapes are cut out.

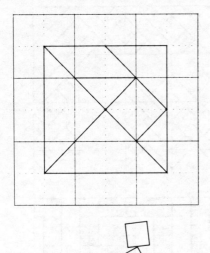

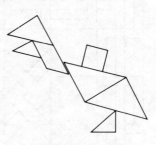

"QUACK! QUACK!"

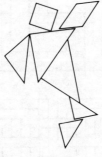

"WAIT FOR ME!"   "GOAL!"

27

Making Designs

Here is another opportunity to make use of shapes. The following examples show how patterns can be made by drawing lines and filling in the squares on graph paper. Once you have decided on a pattern, you must count the squares very carefully in order to repeat the pattern correctly. You may copy some of these suggestions or make up some of your own. The use of coloured pencils will make the designs look even more attractive.

5 How many? – Multiplication

Save time in calculations by learning all of these multiplications. You will not always have your tables with you.

$$0 \times 2 = 2 \times 0 = 0$$
$$1 \times 2 = 2 \times 1 = 2$$
$$2 \times 2 = 2 \times 2 = 4$$
$$3 \times 2 = 2 \times 3 = 6$$
$$4 \times 2 = 2 \times 4 = 8$$
$$5 \times 2 = 2 \times 5 = 10$$
$$6 \times 2 = 2 \times 6 = 12$$
$$7 \times 2 = 2 \times 7 = 14$$
$$8 \times 2 = 2 \times 8 = 16$$
$$9 \times 2 = 2 \times 9 = 18$$
$$10 \times 2 = 2 \times 10 = 20$$
$$11 \times 2 = 2 \times 11 = 22$$
$$12 \times 2 = 2 \times 12 = 24$$

Exercise 20

Cover the '2 times' table and answer these questions:

1. $2 \times 3 =$	2. $4 \times 2 =$	3. $2 \times 1 =$	4. $0 \times 2 =$
5. $? \times 2 = 8$	6. $2 \times ? = 16$	7. $2 \times 5 =$	8. $6 \times 2 =$
9. $5 \times 2 =$	10. $2 \times 7 =$	11. $2 \times ? = 10$	12. $? \times 2 = 6$
13. $3 \times 2 =$	14. $2 \times 4 =$	15. $2 \times 6 =$	16. $2 \times ? = 14$
17. $2 \times ? = 0$	18. $? \times 2 = 2$	19. $7 \times 2 =$	20. $? \times 2 = 12$
21. $2 \times 8 =$	22. $9 \times 2 =$	23. $? \times 2 = 20$	24. $2 \times ? = 24$
25. $? \times 9 = 18$	26. $? \times 1 = 2$	27. $? \times 3 = 6$	28. $? \times 5 = 10$
29. $? \times 2 = 16$	30. $? \times 2 = 22$	31. $8 \times 2 =$	32. $2 \times 10 =$
33. $2 \times ? = 24$	34. $? \times 2 = 18$	35. $2 \times 11 =$	36. $12 \times 2 =$
37. $? \times 10 = 20$	38. $? \times 8 = 16$	39. $? \times 12 = 24$	40. $? \times 11 = 22$
41. $2 \times 12 =$	42. $11 \times 2 =$	43. $10 \times 2 =$	44. $2 \times ? = 20$
45. $8 \times ? = 16$	46. $12 \times ? = 24$	47. $2 \times ? = 18$	48. $2 \times ? = 12$

Save time and learn these multiplications

$$0 \times 3 = 3 \times 0 = 0$$
$$1 \times 3 = 3 \times 1 = 3$$
$$2 \times 3 = 3 \times 2 = 6$$
$$3 \times 3 = 3 \times 3 = 9$$
$$4 \times 3 = 3 \times 4 = 12$$
$$5 \times 3 = 3 \times 5 = 15$$
$$6 \times 3 = 3 \times 6 = 18$$
$$7 \times 3 = 3 \times 7 = 21$$
$$8 \times 3 = 3 \times 8 = 24$$
$$9 \times 3 = 3 \times 9 = 27$$
$$10 \times 3 = 3 \times 10 = 30$$
$$11 \times 3 = 3 \times 11 = 33$$
$$12 \times 3 = 3 \times 12 = 36$$

Exercise 21

Cover the '3 times' table and answer these questions:

1. $1 \times 3 =$	2. $? \times 3 = 0$	3. $3 \times ? = 6$	4. $3 \times 3 =$
5. $3 \times ? = 3$	6. $3 \times 2 =$	7. $4 \times 3 =$	8. $? \times 3 = 6$
9. $3 \times 5 =$	10. $3 \times ? = 12$	11. $? \times 3 = 21$	12. $6 \times 3 =$
13. $3 \times ? = 33$	14. $10 \times 3 =$	15. $? \times 12 = 36$	16. $? \times 3 = 27$
17. $12 \times 3 =$	18. $3 \times 8 =$	19. $5 \times 3 =$	20. $3 \times ? = 18$
21. $3 \times 0 =$	22. $? \times 3 = 3$	23. $? \times 3 = 12$	24. $3 \times 10 =$
25. $3 \times 4 =$	26. $2 \times 3 =$	27. $3 \times ? = 15$	28. $7 \times 3 =$
29. $4 \times ? = 12$	30. $3 \times 6 =$	31. $? \times 7 = 21$	32. $8 \times 3 =$
33. $? \times 3 = 18$	34. $3 \times ? = 24$	35. $3 \times 9 =$	36. $? \times 3 = 15$
37. $3 \times 7 =$	38. $3 \times ? = 27$	39. $3 \times ? = 30$	40. $3 \times 11 =$
41. $3 \times 12 =$	42. $? \times 3 = 36$	43. $8 \times ? = 24$	44. $3 \times ? = 21$
45. $? \times 3 = 24$	46. $11 \times 3 =$	47. $9 \times ? = 27$	48. $? \times 3 = 30$

Save time and learn these multiplications

$0 \times 4 = 4 \times 0 = 0$
$1 \times 4 = 4 \times 1 = 4$
$2 \times 4 = 4 \times 2 = 8$
$3 \times 4 = 4 \times 3 = 12$

**You should
know these
already.**

$4 \times 4 = 4 \times 4 = 16$
$5 \times 4 = 4 \times 5 = 20$
$6 \times 4 = 4 \times 6 = 24$
$7 \times 4 = 4 \times 7 = 28$
$8 \times 4 = 4 \times 8 = 32$
$9 \times 4 = 4 \times 9 = 36$
$10 \times 4 = 4 \times 10 = 40$
$11 \times 4 = 4 \times 11 = 44$
$12 \times 4 = 4 \times 12 = 48$

Exercise 22

Cover the '4 times' table and answer these questions:

1. $4 \times 1 =$	2. $? \times 4 = 8$	3. $4 \times 0 =$	4. $4 \times ? = 16$
5. $5 \times 4 =$	6. $4 \times ? = 28$	7. $4 \times 6 =$	8. $? \times 4 = 12$
9. $4 \times 3 =$	10. $8 \times 4 =$	11. $4 \times ? = 40$	12. $9 \times 4 =$
13. $11 \times 4 =$	14. $1 \times 4 =$	15. $4 \times 4 =$	16. $4 \times 5 =$
17. $4 \times ? = 8$	18. $10 \times 4 =$	19. $? \times 4 = 28$	20. $8 \times ? = 32$
21. $? \times 9 = 36$	22. $6 \times 4 =$	23. $? \times 4 = 48$	24. $4 \times 11 =$
25. $? \times 5 = 20$	26. $4 \times 8 =$	27. $4 \times ? = 36$	28. $2 \times 4 =$
29. $4 \times 7 =$	30. $? \times 4 = 4$	31. $3 \times 4 =$	32. $4 \times 10 =$
33. $0 \times 4 =$	34. $4 \times ? = 44$	35. $12 \times 4 =$	36. $4 \times ? = 48$
37. $? \times 6 = 24$	38. $4 \times ? = 12$	39. $? \times 4 = 16$	40. $7 \times 4 =$
41. $? \times 4 = 40$	42. $? \times 4 = 24$	43. $4 \times 2 =$	44. $? \times 4 = 20$
45. $4 \times ? = 4$	46. $4 \times 9 =$	47. $? \times 4 = 0$	48. $4 \times ? = 32$

6 How many? – Algebra

We know that a + 2a + 3a = 6a

> 6a means 6 things called 'a'
> 4b means 4 things called 'b'
> c means 1 thing called 'c'
> 2d means 2 things called 'd'

> If **a** is equal to **2**, then **6a** would be equal to **12**
> If **b** is equal to **4**, then **4b** would be equal to **16**
> If **c** is equal to **3**, then **1c** would still equal **3**
> If **d** is equal to **1**, then **2d** would be equal to **2**

> $$a + c = 2 + 3 = 5; b + d = 4 + 1 = 5$$

Exercise 23

If a = 2, b = 4, c = 3, d = 1, find the values of these:

1. 2a	2. 2b	3. 2c	4. 3d	5. 3a
6. 3b	7. 3c	8. d	9. 4b	10. a
11. 4a	12. 4d	13. 4c	14. b	15. 5d
16. c	17. 5b	18. 5a	19. 12d	20. 10c
21. 8a	22. 6c	23. 7b	24. 5c	25. 9c
26. 7d	27. 9a	28. 11b	29. 8d	30. 6b
31. 2d	32. 12a	33. 10b	34. 7c	35. 6d
36. 6a	37. 10d	38. 11c	39. 8b	40. 12c
41. 8c	42. 10a	43. 9b	44. 11d	45. 12b
46. 7a	47. 9d	48. 11a	49. a + b	50. c + d
51. 2a + c	52. a + 2b	53. 2a + 2b	54. 2b + c	
55. b + 2c	56. 2b + c	57. a + d	58. 2a + d	
59. a + 2d	60. 2a + 2d	61. 2c + d	62. 2b + d	
63. b + 2d	64. c + 2d	65. 2b + 2c	66. 2b + 2d	
67. 2c + 2d	68. a + 2c	69. 2a + 2c	70. a + 3b	
71. 3a + b	72. 3a + 3b	73. 2a + 3b	74. 3a + 2b	
75. a + 3c	76. 3a + c	77. b + 3c	78. 3b + c	
79. 3a + 3c	80. 3b + 3c	81. a + 3d	82. 3a + d	
83. b + 3d	84. 3b + d	85. 3a + 3d	86. 3b + 3d	
87. c + 3d	88. 3c + d	89. 3c + 3d	90. a + 4b	

91. 4a + b	92. b + 4c	93. 4b + c	94. c + 4d
95. 4c + d	96. 4a + 4b	97. 4b + 4c	98. 4c + 4d
99. 4a + 4d	100. 4a + 4b + 4c + 4d		

Exercise 24

If a = 4, b = 3, c = 2, d = 1, find the values of these:

1. a − b	2. a − c	3. a − d	4. b − c
5. b − d	6. c − d	7. c − 2d	8. a − 2c
9. 2a − b	10. 2a − 2b	11. 2a − c	12. 2a − 2c
13. 2a − 3c	14. a − 3d	15. a − 4d	16. 2a − d
17. 2a − 2d	18. 2a − 3d	19. 2a − 4d	20. 2a − 5d
21. 2a − 6d	22. 2a − 7d	23. 2a − 8d	24. 2b − c
25. 2b − 2c	26. 2b − 3c	27. 2b − 2d	28. 2b − 4d
29. 2b − 6d	30. 2c − d	31. 2c − 3d	32. 3a − 4b
33. 5a − 5b	34. 6a − 6b	35. 7a − 7b	36. 8a − 8b
37. 9a − 9b	38. 10a − 10b	39. 11a − 11b	40. 12a − 12b

Exercise 25

Find the value of 'a' in the following:

1. 2a = 8	2. 3a = 6	3. 4a = 4	4. 5a = 10
5. 4a = 8	6. 4a = 20	7. 4a = 32	8. 3a = 3
9. 2a = 18	10. 2a = 2	11. 3a = 27	12. 2a = 14
13. 3a = 9	14. 4a = 24	15. 4a = 36	16. 3a = 12
17. 4a = 12	18. 3a = 15	19. 2a = 4	20. 4a = 44
21. 3a = 18	22. 2a = 22	23. 4a = 40	24. 2a = 16
25. 2a = 6	26. 3a = 21	27. 3a = 30	28. 3a = 24
29. 4a = 16	30. 3a = 33	31. 2a = 10	32. 4a = 48
33. 2a = 12	34. 4a = 28	35. 3a = 36	36. 2a = 24

7 How many? – Easy multiplication

EXAMPLE 1   123
           x3
          369

STEPS: 3 x 3 = **9**
         3 x 2 = **6**
         3 x 1 = 3
         Answer 369

EXAMPLE 2   56p
            x2
          112p

STEPS: 2 x 6 = **12**; write **2** and *'carry 1'*
       2 x 5 = **10**; and the *'carry 1'* makes **11**

Answer 112p or £1.12

Exercise 26

1. 21
  x2

2. 22
  x3

3. 21
  x4

4. 22
  x2

5. 13
  x3

6. 14
  x1

7. 21
  x3

8. 22
  x4

9. 14
  x2

10. 23
  x3

11. 24
  x2

12. 23
  x2

13. 20
  x3

14. 31
  x2

15. 33
  x3

16. 34
  x2

17. 41
  x2

18. 30
  x3

19. 20
  x4

20. 42
  x2

21. 121
  x3

22. 222
  x2

23. 101
  x4

24. 202
  x4

25. 40
  x2

26. 103
  x3

27. 123
  x2

28. 134
  x2

29. 62	30. 132	31. 122	32. 321
x2	x3	x4	x2

33. 123	34. 122	35. 124	36. 222
x3	x2	x2	x4

37. 333	38. 221	39. 423	40. 303
x3	x4	x2	x3

41. 24p	42. 25p	43. 23p	44. 35p
x3	x2	x4	x2
p	p	p	p

45. 25p	46. 24p	47. 28p	48. 27p
x3	x4	x3	x2
p	p	p	p

49. 26p	50. 27p	51. 28p	52. 29p
x2	x2	x2	x2
p	p	p	p

53. 30p	54. 34p	55. 35p	56. 36p
x2	x3	x3	x3
p	p	p	p

57. 37p	58. 38p	59. 39p	60. 40p
x3	x3	x3	x3
p	p	p	p

61. 43p	62. 44p	63. 44p	64. 45p
x4	x4	x5	x4
p	p	p	p

65. 40p	66. 60p	67. 40p	68. 70p
x6	x4	x7	x4
£	£	£	£

Save time and learn these multiplications

```
0 x 5 = 5 x 0 = 0
1 x 5 = 5 x 1 = 5
2 x 5 = 5 x 2 = 10
3 x 5 = 5 x 3 = 15
4 x 5 = 5 x 4 = 20
 5 x 5 = 5 x  5 = 25
 6 x 5 = 5 x  6 = 30
 7 x 5 = 5 x  7 = 35
 8 x 5 = 5 x  8 = 40
 9 x 5 = 5 x  9 = 45
10 x 5 = 5 x 10 = 50
11 x 5 = 5 x 11 = 55
12 x 5 = 5 x 12 = 60
```

**You should
know these
already**

Exercise 27

Cover the '5 times' table and answer these questions:

1. 5 x 1 =	2. ? x 5 = 10	3. 5 x 0 =	4. 5 x ? = 20
5. 5 x 5 =	6. 5 x ? = 35	7. 5 x 6 =	8. ? x 5 = 15
9. 5 x 3 =	10. 8 x 5 =	11. 5 x ? = 50	12. 9 x 5 =
13. 11 x 5 =	14. 1 x 5 =	15. 5 x 5 =	16. 4 x 5 =
17. 5 x ? = 10	18. 10 x 5 =	19. ? x 5 = 35	20. 8 x ? = 40
21. ? x 9 = 45	22. 6 x 5 =	23. ? x 5 = 60	24. 5 x 11 =
25. ? x 5 = 25	26. 5 x 8 =	27. 5 x ? = 45	28. 2 x 5 =
29. 5 x 7 =	30. ? x 5 = 5	31. 3 x 5 =	32. 5 x 10 =
33. 0 x 5 =	34. 5 x ? = 55	35. 12 x 5 =	36. 5 x ? = 60
37. ? x 6 = 30	38. 5 x ? = 15	39. ? x 5 = 20	40. 7 x 5 =
41. ? x 5 = 50	42. ? x 5 = 30	43. 5 x 2 =	44. ? x 5 = 25
45. 5 x ? = 5	46. 5 x 9 =	47. ? x 5 = 0	48. 5 x ? = 40

Save time and learn these multiplications

0 x 6 = 6 x 0 = 0
1 x 6 = 6 x 1 = 6
2 x 6 = 6 x 2 = 12
3 x 6 = 6 x 3 = 18
4 x 6 = 6 x 4 = 24
5 x 6 = 6 x 5 = 30

You should
know these
already

6 x 6 = 6 x 6 = 36
7 x 6 = 6 x 7 = 42
8 x 6 = 6 x 8 = 48
9 x 6 = 6 x 9 = 54
10 x 6 = 6 x 10 = 60
11 x 6 = 6 x 11 = 66
12 x 6 = 6 x 12 = 72

Exercise 28

Cover the '6 times' table and answer these questions:

1.	1 x 6 =	**2.**	? x 6 = 0	**3.**	6 x ? = 12	**4.**	6 x 6 =
5.	6 x ? = 6	**6.**	6 x 2 =	**7.**	4 x 6 =	**8.**	? x 6 = 12
9.	6 x 5 =	**10.**	6 x ? = 24	**11.**	? x 6 = 42	**12.**	6 x 3 =
13.	6 x ? = 66	**14.**	10 x 6 =	**15.**	? x 12 = 72	**16.**	? x 6 = 54
17.	12 x 6 =	**18.**	6 x 8 =	**19.**	5 x 6 =	**20.**	6 x ? = 36
21.	6 x 0 =	**22.**	? x 6 = 6	**23.**	? x 6 = 24	**24.**	6 x 10 =
25.	6 x 4 =	**26.**	2 x 6 =	**27.**	6 x ? = 30	**28.**	7 x 6 =
29.	4 x ? = 24	**30.**	3 x 6 =	**31.**	? x 7 = 42	**32.**	8 x 6 =
33.	? x 6 = 36	**34.**	6 x ? = 24	**35.**	6 x 9 =	**36.**	? x 6 = 30
37.	6 x 7 =	**38.**	6 x ? = 54	**39.**	6 x ? = 60	**40.**	6 x 11 =
41.	6 x 12 =	**42.**	? x 6 = 72	**43.**	8 x ? = 48	**44.**	6 x ? = 42
45.	? x 6 = 48	**46.**	11 x 6 =	**47.**	9 x ? = 54	**48.**	? x 6 = 60

Save time and learn these multiplications

0 x 7 = 7 x 0 = 0
1 x 7 = 7 x 1 = 7
2 x 7 = 7 x 2 = 14
3 x 7 = 7 x 3 = 21
4 x 7 = 7 x 4 = 28
5 x 7 = 7 x 5 = 35
6 x 7 = 7 x 6 = 42
7 x 7 = 7 x 7 = 49
8 x 7 = 7 x 8 = 56
9 x 7 = 7 x 9 = 63
10 x 7 = 7 x 10 = 70
11 x 7 = 7 x 11 = 77
12 x 7 = 7 x 12 = 84

You should
know these
already

Exercise 29

Cover the '7 times' table and answer these questions:

1. 7 x 1 =	2. ? x 7 = 14	3. 7 x 0 =	4. 7 x ? = 49
5. 5 x 7 =	6. 7 x ? = 28	7. 7 x 6 =	8. ? x 7 = 21
9. 7 x 3 =	10. 8 x 7 =	11. 7 x ? = 70	12. 9 x 7 =
13. 11 x 7 =	14. 1 x 7 =	15. 7 x 7 =	16. 7 x 5 =
17. 7 x ? = 14	18. 10 x 7 =	19. ? x 7 = 28	20. 8 x ? = 56
21. ? x 9 = 63	22. 6 x 7 =	23. ? x 7 = 84	24. 7 x 11 =
25. ? x 5 = 35	26. 7 x 8 =	27. 7 x ? = 63	28. 2 x 7 =
29. 4 x 7 =	30. ? x 7 = 7	31. 3 x 7 =	32. 7 x 10 =
33. 0 x 7 =	34. 7 x ? = 77	35. 12 x 7 =	36. 7 x ? = 84
37. ? x 6 = 42	38. 7 x ? = 21	39. ? x 7 = 49	40. 7 x 4 =
41. ? x 7 = 70	42. ? x 7 = 42	43. 7 x 2 =	44. ? x 7 = 35
45. 7 x ? = 7	46. 7 x 9 =	47. ? x 7 = 0	48. 7 x ? = 56

Save time and learn these multiplications

$$0 \times 8 = 8 \times 0 = 0$$
$$1 \times 8 = 8 \times 1 = 8$$
$$2 \times 8 = 8 \times 2 = 16$$
$$3 \times 8 = 8 \times 3 = 24$$
$$4 \times 8 = 8 \times 4 = 32$$
$$5 \times 8 = 8 \times 5 = 40$$
$$6 \times 8 = 8 \times 6 = 48$$
$$7 \times 8 = 8 \times 7 = 56$$
$$8 \times 8 = 8 \times \ 8 = 64$$
$$9 \times 8 = 8 \times \ 9 = 72$$
$$10 \times 8 = 8 \times 10 = 80$$
$$11 \times 8 = 8 \times 11 = 88$$
$$12 \times 8 = 8 \times 12 = 96$$

You should
know these
already

Exercise 30

Cover the '8 times' table and answer these questions:

1.	$1 \times 8 =$	**2.**	$? \times 8 = 0$	**3.**	$8 \times ? = 16$	**4.**	$8 \times 8 =$
5.	$8 \times ? = 8$	**6.**	$8 \times 2 =$	**7.**	$4 \times 8 =$	**8.**	$? \times 8 = 16$
9.	$8 \times 5 =$	**10.**	$8 \times ? = 32$	**11.**	$? \times 8 = 56$	**12.**	$8 \times 3 =$
13.	$8 \times ? = 88$	**14.**	$10 \times 8 =$	**15.**	$? \times 12 = 96$	**16.**	$? \times 8 = 72$
17.	$12 \times 8 =$	**18.**	$6 \times 8 =$	**19.**	$5 \times 8 =$	**20.**	$8 \times ? = 64$
21.	$8 \times 0 =$	**22.**	$? \times 8 = 8$	**23.**	$? \times 8 = 32$	**24.**	$8 \times 10 =$
25.	$8 \times 4 =$	**26.**	$2 \times 8 =$	**27.**	$8 \times ? = 40$	**28.**	$7 \times 8 =$
29.	$4 \times ? = 32$	**30.**	$3 \times 8 =$	**31.**	$? \times 7 = 56$	**32.**	$8 \times 6 =$
33.	$? \times 8 = 64$	**34.**	$8 \times ? = 32$	**35.**	$8 \times 9 =$	**36.**	$? \times 8 = 40$
37.	$8 \times 7 =$	**38.**	$8 \times ? = 72$	**39.**	$8 \times ? = 80$	**40.**	$8 \times 11 =$
41.	$8 \times 12 =$	**42.**	$? \times 8 = 96$	**43.**	$8 \times ? = 48$	**44.**	$8 \times ? = 56$
45.	$? \times 8 = 48$	**46.**	$11 \times 8 =$	**47.**	$9 \times ? = 72$	**48.**	$? \times 8 = 80$

8 How many? – Algebra

3a means 3 things called 'a'
b means 1 thing called 'b'
2c means 2 things called 'c'
5d means 5 things called 'd'

If **a** is equal to **7**, then **3a** would be equal to **21**
If **b** is equal to **6**, then **1b** would still equal **6**
If **c** is equal to **8**, then **2c** would be equal to **16**
If **d** is equal to **5**, then **5d** would be equal to **25**

$$a + b = 7 + 6 = 13; c - d = 8 - 5 = 3$$

Exercise 31

If $a = 7$, $b = 6$, $c = 8$, $d = 5$, find the values of these:

1. 2a	2. 2b	3. 2c	4. 3d	5. 3a
6. 3b	7. 3c	8. d	9. 4b	10. a
11. 5d	12. b	13. 4c	14. 4d	15. 4a
16. 10c	17. 12d	18. 5a	19. 5b	20. c
21. 9c	22. 5c	23. 7b	24. 6c	25. 8a
26. 6b	27. 8d	28. 11b	29. 9a	30. 7d
31. 6d	32. 7c	33. 10b	34. 12a	35. 2d
36. 12c	37. 8b	38. 11c	39. 10d	40. 6a
41. 12b	42. 11d	43. 9b	44. 10a	45. 8c
46. 11a	47. 9d	48. 7a	49. a + c	50. b + d
51. 2a + c	52. a + 2b	53. 2a + 2b		54. b + c
55. 2a + d	56. a + d	57. 2b + c		58. b + 2c
59. 2b + d	60. 2c + d	61. 2a + 2d		62. a + 2d
63. 2b + 2d	64. 2b + 2c	65. c + 2d		66. b + 2d
67. 2c + 2d	68. a + 2c	69. 2a + 2c		70. a + 3b
71. 3a + b	72. 3a + 3b	73. 2a + 3b		74. 3a + 2b

75. a + 3c	76. 3a + c	77. b + 3c	78. 3b + c
79. 3a + 3c	80. 3b + 3c	81. a + 3d	82. 3a + d
83. b + 3d	84. 3b + d	85. 3a + 3d	86. 3b + 3d
87. c + 3d	88. 3c + d	89. 3c + 3d	90. a + 4b
91. 4a + b	92. b + 4c	93. 4b + c	94. c + 4d
95. 4c + d	96. 4a + 4b	97. 4b + 4c	98. 4c + 4d
99. 4a + 4d	100. 4a + 4b + 4c + 4d.		

Exercise 32

If **a = 8, b = 7, c = 6, d = 5,** find the values of these:

1. a − b	2. a − c	3. a − d	4. b − c
5. b − d	6. c − d	7. 2c − 2d	8. 2a − 2c
9. 2a − b	10. 2a − 2b	11. 2a − c	12. 3a − 2c
13. 3a − 3c	14. 3d − a	15. 4d − a	16. 2a − d
17. 2a − 2d	18. 2a − 3d	19. 4d − 2a	20. 5d − 2a
21. 6d − 2a	22. 7d − 2a	23. 8d − 2a	24. 2b − c
25. 2b − 2c	26. 3c − 2b	27. 2b − 2d	28. 4d − 2b
29. 6d − 2b	30. 2c − d	31. 3d − 2c	32. 4b − 3a
33. 5a − 5b	34. 6a − 6b	35. 7a − 7b	36. 8a − 8b
37. 9a − 9b	38. 10a − 10b	39. 11a − 11b	40. 12a − 12b

Exercise 33

Find the value of 'a' in the following:

1. 5a = 45	2. 6a = 0	3. 2a = 14	4. 8a = 8
5. 7a = 7	6. 2a = 16	7. 8a = 0	8. 5a = 30
9. 3a = 18	10. 11a = 55	11. 7a = 0	12. 3a = 24
13. 4a = 20	14. 8a = 32	15. 8a = 48	16. 6a = 6
17. 8a = 40	18. 3a = 21	19. 5a = 40	20. 7a = 56
21. 2a = 12	22. 11a = 66	23. 8a = 64	24. 7a = 28
25. 3a = 15	26. 9a = 72	27. 6a = 24	28. 5a = 25
29. 5a = 30	30. 6a = 72	31. 5a = 5	32. 6a = 36
33. 10a = 50	34. 5a = 35	35. 6a = 42	36. 7a = 49
37. 11a = 88	38. 8a = 56	39. 7a = 63	40. 5a = 10
41. 7a = 35	42. 7a = 42	43. 10a = 70	44. 8a = 80
45. 12a = 96	46. 11a = 77	47. 5a = 60	48. 6a = 48
49. 9a = 54	50. 5a = 0	51. 10a = 60	52. 12a = 84

9 How many? – More easy multiplication

STEPS:

EXAMPLE 1   123p
         x6
Answer 738p
or £7.38

$6 \times 3 = 18$; write **8** and *'carry 1'*
$6 \times 2 = 12$; and the *'carry 1'* makes 13
         write **3** and *'carry 1'*
$6 \times 1 = 6$; and the *'carry 1'* makes **7** write **7**

STEPS:

EXAMPLE 2   746
         x8
Answer 5968

$8 \times 6 = 48$; write **8**, *'carry 4'*
$8 \times 4 = 32$; and the *'carry 4'* = **36**
         write **6**, *'carry 3'*
$8 \times 7 = 56$; and the *'carry 3'* = **59** write **59**

Exercise 34

1. 22 x6	**2.** 33 x5	**3.** 14 x8	**4.** 15 x9
5. 17 x5	**6.** 18 x6	**7.** 13 x9	**8.** 24 x8
9. 16 x7	**10.** 23 x6	**11.** 33 x8	**12.** 27 x5
13. 37 x7	**14.** 45 x7	**15.** 36 x8	**16.** 45 x6
17. 28 x5	**18.** 29 x6	**19.** 18 x8	**20.** 19 x7

21. 32 x8	22. 53 x7	23. 64 x6	24. 78 x5

25. 123 x5	26. 132 x6	27. 143 x7	28. 154 x8

29. 268 x5	30. 145 x6	31. 215 x7	32. 223 x8

33. 176 x5	34. 278 x6	35. 187 x7	36. 268 x8

37. 24p x5 ___ p	38. 25p x6 ___ p	39. 22p x7 ___ p	40. 23p x8 ___ p

41. 18p x8 ___ p	42. 19p x7 ___ p	43. 16p x6 ___ p	44. 17p x5 ___ p

45. 34p x5 ___ p	46. 30p x6 ___ p	47. 35p x7 ___ p	48. 36p x8 ___ p

49. 45p x8 ___ p	50. 56p x7 ___ p	51. 67p x6 ___ p	52. 78p x4 ___ p

53. 31p x5 £ ___	54. 35p x6 £ ___	55. 37p x8 £ ___	56. 38p x7 £ ___

57. 42p x8 £ ___	58. 56p x7 £ ___	59. 67p x6 £ ___	60. 78p x5 £ ___

61. 36p x5 £ ___	62. 45p x6 £ ___	63. 54p x7 £ ___	64. 63p x8 £ ___

10 The circle – More designs

Circles are drawn with the aid of **compasses**. There are several words used to name different parts of the circle but at the moment we need to know only four of them:

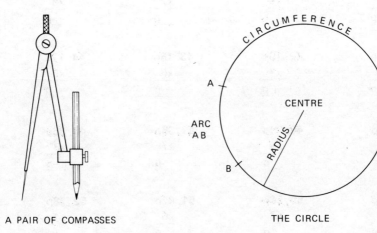

A PAIR OF COMPASSES   THE CIRCLE

The **centre** is the position where the **point** of the compass is placed.
The **radius** is the **distance between** the compass **point** and the compass **pencil**.
The **circumference** is the path traced out by the moving point of the pencil.
An **arc** is any part of the circumference of a circle — a curved line.
Note. The plural form of radius (when there are several) is called **radii**.

Using the compasses

1. Open the compasses and draw a circle.

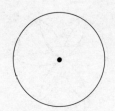

2. Keeping the same radius, set the point of the compasses on the circumference and draw an arc from one side of the circumference to the other. It should pass through the centre.

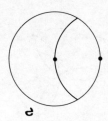

3. Still with the same radius, move the compasses to each of the points where the arc has cut the circumference, drawing a new arc each time.

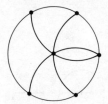

4. Repeat the process until you have obtained the completed design. Notice that the figure has six leaves, this is because the **radius can be stepped off six times round the circumference.** This is always so, no matter what the size of the circle.

The diagrams overleaf show various methods of drawing circle designs.

C shows how a polygon may be obtained by joining the six points on the circumference. What is the name of such a polygon?

D shows the result of using alternate arcs only.

G and H use smaller circles inside the large one.

I and J also make use of smaller circles (half the radius of the big one) and they also show the effect of shading.

K and L are interlocking designs and need great care. (Over one and under the next.)

Copy these designs and try to draw some others of your own.

A B C

D E F

G H I

J K L

11 How many? – Multiplication

Save time and learn these multiplications

$$0 \times 9 = 9 \times 0 = 0$$
$$1 \times 9 = 9 \times 1 = 9$$
$$2 \times 9 = 9 \times 2 = 18$$
$$3 \times 9 = 9 \times 3 = 27$$
$$4 \times 9 = 9 \times 4 = 36$$
$$5 \times 9 = 9 \times 5 = 45$$
$$6 \times 9 = 9 \times 6 = 54$$
$$7 \times 9 = 9 \times 7 = 63$$
$$8 \times 9 = 9 \times 8 = 72$$
$$9 \times 9 = 9 \times \ 9 = 81$$
$$10 \times 9 = 9 \times 10 = 90$$
$$11 \times 9 = 9 \times 11 = 99$$
$$12 \times 9 = 9 \times 12 = 108$$

You should
know these
already

Exercise 35

Cover the '9 times' table and answer these questions.

1. $4 \times 9 =$	**2.** $0 \times 9 =$	**3.** $9 \times 7 =$	**4.** $9 \times 2 =$
5. $? \times 9 = 9$	**6.** $8 \times 9 =$	**7.** $10 \times 9 =$	**8.** $9 \times ? = 54$
9. $9 \times ? = 108$	**10.** $? \times 9 = 90$	**11.** $? \times 9 = 36$	**12.** $? \times 9 = 27$
13. $9 \times 6 =$	**14.** $9 \times 1 =$	**15.** $9 \times 9 =$	**16.** $9 \times ? = 63$
17. $9 \times ? = 0$	**18.** $11 \times 9 =$	**19.** $? \times 9 = 18$	**20.** $? \times 9 = 72$
21. $12 \times 9 =$	**22.** $? \times 9 = 99$	**23.** $9 \times 12 =$	**24.** $? \times 9 = 108$
25. $? \times 9 = 45$	**26.** $9 \times 3 =$	**27.** $? \times 9 = 63$	**28.** $9 \times ? = 9$
29. $9 \times ? = 99$	**30.** $? \times 9 = 81$	**31.** $9 \times 0 =$	**32.** $9 \times 5 =$
33. $9 \times ? = 18$	**34.** $9 \times ? = 90$	**35.** $? \times 9 = 54$	**36.** $9 \times ? = 81$
37. $9 \times 8 =$	**38.** $9 \times ? = 36$	**39.** $9 \times ? = 45$	**40.** $9 \times ? = 27$
41. $1 \times 9 =$	**42.** $9 \times 11 =$	**43.** $9 \times 10 =$	**44.** $7 \times 9 =$
45. $6 \times 9 =$	**46.** $2 \times 9 =$	**47.** $9 \times 4 =$	**48.** $? \times 9 = 0$
49. $3 \times 9 =$	**50.** $9 \times ? = 72$	**51.** $5 \times 9 =$	

Save time and learn these multiplications

**You should
know these
already**

10 x 10 = 10 x 10 = 100   The rest
11 x 10 = 10 x 11 = 110   are easy
12 x 10 = 10 x 12 = 120

Exercise 36

Cover the '10 times' table and answer these questions:

1. 2 x 10 =	**2.** 10 x ? = 30	**3.** 10 x 4 =	**4.** 1 x 10 =
5. 10 x 8 =	**6.** 10 x 0 =	**7.** ? x 10 = 60	**8.** 10 x ? = 70
9. 10 x ? = 50	**10.** 9 x 10 =	**11.** 10 x 11 =	**12.** ? x 10 = 40
13. ? x 10 = 10	**14.** ? x 10 = 100	**15.** 8 x 10 =	**16.** 7 x 10 =
17. 10 x 10 =	**18.** 10 x 2 =	**19.** ? x 10 = 30	**20.** ? x 10 = 0
21. 10 x ? = 60	**22.** 12 x 10 =	**23.** 10 x ? = 110	**24.** 10 x 9 =
25. 0 x 10 =	**26.** ? x 10 = 50	**27.** 10 x ? = 40	**28.** 10 x ? = 100
29. 10 x 5 =	**30.** 10 x 1 =	**31.** 10 x 7 =	**32.** 3 x 10 =
33. 10 x 12 =	**34.** 10 x ? = 80	**35.** ? x 10 = 20	**36.** 11 x 10 =
37. ? x 10 = 70	**38.** ? x 10 = 90	**39.** 6 x 10 =	**40.** 5 x 10 =
41. 10 x ? = 20	**42.** 4 x 10 =	**43.** ? x 10 = 120	**44.** ? x 10 = 80
45. 10 x 6 =	**46.** ? x 10 = 110	**47.** 10 x ? = 0	**48.** 10 x ? = 90
49. 10 x ? = 10	**50.** 10 x 3 =	**51.** 10 x ? = 120	

Save time and learn these multiplications

$$0 \times 11 = 11 \times 0 = 0$$
$$1 \times 11 = 11 \times 1 = 11$$
$$2 \times 11 = 11 \times 2 = 22$$
$$3 \times 11 = 11 \times 3 = 33$$
$$4 \times 11 = 11 \times 4 = 44$$
$$5 \times 11 = 11 \times 5 = 55$$
$$6 \times 11 = 11 \times 6 = 66$$
$$7 \times 11 = 11 \times 7 = 77$$
$$8 \times 11 = 11 \times 8 = 88$$
$$9 \times 11 = 11 \times 9 = 99$$
$$10 \times 11 = 11 \times 10 = 110$$

**You should
know these
already**

$$11 \times 11 = 11 \times 11 = 121$$
$$12 \times 11 = 11 \times 12 = 132$$

Just add **11**
each time

Exercise 37

Cover the '11 times' table and answer these questions:

1. $11 \times ? = 22$	**2.** $11 \times 5 =$	**3.** $1 \times 11 =$	**4.** $11 \times ? = 66$
5. $7 \times 11 =$	**6.** $? \times 11 = 33$	**7.** $11 \times 8 =$	**8.** $4 \times 11 =$
9. $11 \times 0 =$	**10.** $9 \times 11 =$	**11.** $11 \times ? = 99$	**12.** $10 \times 11 =$
13. $? \times 11 = 66$	**14.** $11 \times 1 =$	**15.** $11 \times 7 =$	**16.** $? \times 11 = 55$
17. $11 \times ? = 121$	**18.** $? \times 11 = 110$	**19.** $? \times 11 = 22$	**20.** $11 \times ? = 33$
21. $? \times 11 = 44$	**22.** $0 \times 11 =$	**23.** $11 \times ? = 88$	**24.** $? \times 11 = 132$
25. $11 \times ? = 110$	**26.** $11 \times 12 =$	**27.** $11 \times 4 =$	**28.** $11 \times 9 =$
29. $3 \times 11 =$	**30.** $11 \times ? = 55$	**31.** $11 \times 6 =$	**32.** $? \times 11 = 77$
33. $12 \times 11 =$	**34.** $? \times 11 = 88$	**35.** $11 \times ? = 0$	**36.** $11 \times 2 =$
37. $? \times 11 = 11$	**38.** $11 \times 3 =$	**39.** $11 \times 11 =$	**40.** $11 \times ? = 132$
41. $? \times 11 = 99$	**42.** $? \times 11 = 121$	**43.** $11 \times ? = 77$	**44.** $5 \times 11 =$
45. $11 \times 10 =$	**46.** $11 \times ? = 11$	**47.** $8 \times 11 =$	**48.** $? \times 11 = 0$
49. $11 \times ? = 44$	**50.** $6 \times 11 =$	**51.** $2 \times 11 =$	

Save time and learn these multiplications

$$0 \times 12 = 12 \times 0 = 0$$
$$1 \times 12 = 12 \times 1 = 12$$
$$2 \times 12 = 12 \times 2 = 24$$
$$3 \times 12 = 12 \times 3 = 36$$
$$4 \times 12 = 12 \times 4 = 48$$
$$5 \times 12 = 12 \times 5 = 60$$
$$6 \times 12 = 12 \times 6 = 72$$
$$7 \times 12 = 12 \times 7 = 84$$
$$8 \times 12 = 12 \times 8 = 96$$
$$9 \times 12 = 12 \times 9 = 108$$
$$10 \times 12 = 12 \times 10 = 120$$
$$11 \times 12 = 12 \times 11 = 132$$

You should
know these
already

$$12 \times 12 = 12 \times 12 = 144$$

Only one left

Exercise 38

Cover the '12 times' table and answer these questions:

1. $? \times 12 = 24$	2. $4 \times 12 =$	3. $12 \times 7 =$	4. $? \times 12 = 96$
5. $9 \times 12 =$	6. $? \times 12 = 120$	7. $5 \times 12 =$	8. $12 \times ? = 12$
9. $12 \times ? = 36$	10. $0 \times 12 =$	11. $12 \times 6 =$	12. $10 \times 12 =$
13. $11 \times 12 =$	14. $12 \times ? = 132$	15. $12 \times ? = 24$	16. $? \times 12 = 108$
17. $12 \times ? = 144$	18. $? \times 12 = 12$	19. $? \times 12 = 72$	20. $7 \times 12 =$
21. $? \times 12 = 0$	22. $12 \times 9 =$	23. $12 \times 4 =$	24. $? \times 12 = 36$
25. $8 \times 12 =$	26. $? \times 12 = 48$	27. $? \times 12 = 132$	28. $12 \times 5 =$
29. $12 \times ? = 60$	30. $12 \times 12 =$	31. $12 \times ? = 96$	32. $? \times 12 = 144$
33. $12 \times 1 =$	34. $12 \times 2 =$	35. $3 \times 12 =$	36. $12 \times ? = 0$
37. $? \times 12 = 84$	38. $12 \times ? = 72$	39. $12 \times 8 =$	40. $12 \times 10 =$
41. $6 \times 12 =$	42. $12 \times ? = 48$	43. $? \times 12 = 60$	44. $1 \times 12 =$
45. $2 \times 12 =$	46. $12 \times 11 =$	47. $12 \times ? = 120$	48. $12 \times ? = 84$
49. $12 \times ? = 108$	50. $12 \times 0 =$	51. $12 \times 3 =$	

12 How many? – Algebra

2a means 2 things called 'a'
3b means 3 things called 'b'
c means 1 thing called 'c'
4d means 4 things called 'd'

If **a** is equal to **9**, then **2a** would be equal to **18**
If **b** is equal to **10**, then **3b** would be equal to **30**
If **c** is equal to **11**, then **1c** would still equal **11**
If **d** is equal to **12**, then **4d** would be equal to **48**

$$a + c = 9 + 11 = 20; d - b = 12 - 10 = 2$$

Exercise 39

If $a = 9, b = 10, c = 11, d = 12$, find the values of these:

1.	5b	2.	a	3.	11c	4.	4d	5.	11d
6.	12d	7.	8d	8.	5c	9.	b	10.	7a
11.	4b	12.	11a	13.	3a	14.	10d	15.	bc
16.	2a	17.	3d	18.	12b	19.	2c	20.	9b
21.	4c	22.	2b	23.	12c	24.	7b	25.	8a
26.	3b	27.	5a	28.	6b	29.	4a	30.	10c
31.	7d	32.	3c	33.	12a	34.	9c	35.	6d
36.	6a	37.	11b	38.	8c	39.	d	40.	7c
41.	9d	42.	2d	43.	10b	44.	9a	45.	8b
46.	10a	47.	c	48.	5d	49.	a + b	50.	a + d
51.	2a + c	52.	a + 2b	53.	2a + 2b	54.	b + c		
55.	b + 2c	56.	2b + c	57.	a + d	58.	2a + d		
59.	a + 2d	60.	2a + 2d	61.	2c + d	62.	2b + d		
63.	b + 2d	64.	c + 2d	65.	2b + 2c	66.	2b + 2d		
67.	2c + 2d	68.	a + 2c	69.	2a + 2c	70.	a + 3b		
71.	3a + b	72.	3a + 3b	73.	2a + 3b	74.	3a + 2b		
75.	a + 3c	76.	3a + c	77.	b + 3c	78.	3b + c		
79.	3a + 3c	80.	3b + 3c	81.	a + 3d	82.	3a + d		
83.	b + 3d	84.	3b + d	85.	3a + 3d	86.	3b + 3d		
87.	c + 3d	88.	3c + d	89.	3c + 3d	90.	a + 4b		

Exercise 40

If **a = 12, b = 11, c = 10, d = 9,** find the values of these:

1. $a - b$	2. $a - c$	3. $a - d$	4. $b - c$
5. $b - d$	6. $c - d$	7. $2c - d$	8. $2a - c$
9. $2a - b$	10. $2a - 2b$	11. $2a - 2c$	12. $3a - 2c$
13. $3a - 3c$	14. $3d - a$	15. $4d - a$	16. $2a - d$
17. $2a - 2d$	18. $3a - 2d$	19. $4d - 2a$	20. $5d - 2a$
21. $6d - 2a$	22. $7d - 2a$	23. $8d - 2a$	24. $2b - c$
25. $2b - 2c$	26. $3c - 2b$	27. $2b - 2d$	28. $4d - 2b$
29. $6d - 2b$	30. $2c - d$	31. $3d - 2c$	32. $4b - 3a$
33. $5a - 5b$	34. $6a - 6b$	35. $7a - 7b$	36. $8a - 8b$
37. $9a - 9b$	38. $10a - 10b$	39. $11a - 11b$	40. $12a - 12b$

Exercise 41

Find the value of **'a'** in the following:

1. $10a = 30$	2. $9a = 27$	3. $12a = 120$	4. $11a = 99$
5. $9a = 99$	6. $11a = 132$	7. $10a = 80$	8. $9a = 45$
9. $12a = 72$	10. $10a = 40$	11. $9a = 18$	12. $11a = 77$
13. $10a = 110$	14. $9a = 108$	15. $11a = 88$	16. $12a = 144$
17. $12a = 96$	18. $10a = 120$	19. $11a = 66$	20. $9a - 90$
21. $9a = 9$	22. $11a = 121$	23. $10a = 50$	24. $12a = 132$
25. $12a = 108$	26. $9a = 0$	27. $12a = 84$	28. $11a = 55$
29. $10a = 70$	30. $12a = 36$	31. $9a = 54$	32. $10a = 20$
33. $12a = 60$	34. $11a = 44$	35. $10a = 100$	36. $11a = 0$
37. $9a = 36$	38. $10a = 10$	39. $12a = 24$	40. $9a = 63$
41. $11a = 22$	42. $9a = 72$	43. $10a = 90$	44. $12a = 12$
45. $11a = 110$	46. $12a = 0$	47. $11a = 33$	48. $10a = 60$
49. $10a = 0$	50. $11a = 11$	51. $9a = 81$	52. $12a = 48$

13 How many? – Larger Numbers

Multiplying by 10, 100, 1000

From the 'table of tens' and the '10 times table' on page 48 we see such facts as:

$$3 \times 10 = 10 \times 3 = 30$$
$$5 \times 10 = 10 \times 5 = 50$$

These are easy to learn and do not need to be worked out.
We just put a nought (0) after the figure we are using:

$$3 \times 10 \rightarrow 3 \text{ becomes } 30$$
$$5 \times 10 \rightarrow 5 \text{ becomes } 50$$

If we used a '100 times table', **we would put two noughts (00) after the figure we are using:**

$$3 \times 100 \rightarrow 3 \text{ becomes } 300$$
$$5 \times 100 \rightarrow 5 \text{ becomes } 500$$

If we used a '1000 times table', **we would put three noughts (000) after the figure we are using:**

$$3 \times 1000 \rightarrow 3 \text{ becomes } 3000$$
$$5 \times 1000 \rightarrow 5 \text{ becomes } 5000$$

Exercise 42

1. 2 x 100	2. 4 x 1000	3. 13 x 10	4. 12 x 10
5. 1 x 1000	6. 8 x 10	7. 12 x 100	8. 7 x 1000
9. 24 x 10	10. 1 x 100	11. 57 x 100	12. 6 x 100
13. 5 x 10	14. 12 x 1000	15. 3 x 10	16. 68 x 100
17. 8 x 1000	18. 7 x 100	19. 79 x 100	20. 5 x 1000
21. 11 x 100	22. 35 x 10	23. 3 x 1000	24. 9 x 10
25. 24 x 1000	26. 4 x 100	27. 8 x 100	28. 35 x 1000
29. 1 x 10	30. 11 x 1000	31. 9 x 1000	32. 46 x 1000
33. 2 x 1000	34. 46 x 10	35. 7 x 10	36. 10 x 100
37. 3 x 100	38. 2 x 10	39. 9 x 100	40. 11 x 10
41. 58 x 1000	42. 6 x 1000	43. 69 x 100	44. 71 x 10
45. 6 x 10	46. 83 x 1000	47. 5 x 100	48. 10 x 1000

Multiplying by 20, 30, 40, etc

EXAMPLE 1   20 x 12

	20
	x12
Answer	240

Steps:
12 times 0 = 0, write down the '**0**'
12 times 2 = 24, write down the '**24**'

EXAMPLE 2   12 x 20

	12
	x20
Answer	240

Steps:
Add a '**0**' for the '*times 10*', write down the '**0**'
2 times 12 = 24, write down the '**24**'

RESULT: **20 x 12 = 12 x 20 = 240**

NOTE   (i) 20, 30, 40, etc. are called '**multiples of ten**', that is '**groups of tens**'.
(ii) To multiply by '*multiples of ten*', **add a nought** for the tens **then multiply by the number in front.**

Exercise 43

1.	12 x 30	**2.**	12 x 40	**3.**	12 x 50	**4.**	12 x 60
5.	12 x 70	**6.**	12 x 80	**7.**	12 x 90	**8.**	6 x 20
9.	6 x 40	**10.**	6 x 60	**11.**	6 x 80	**12.**	7 x 30
13.	7 x 50	**14.**	7 x 70	**15.**	7 x 90	**16.**	4 x 20
17.	4 x 40	**18.**	4 x 60	**19.**	4 x 80	**20.**	5 x 30
21.	5 x 50	**22.**	5 x 70	**23.**	5 x 90	**24.**	3 x 20
25.	3 x 40	**26.**	3 x 60	**27.**	3 x 80	**28.**	11 x 30
29.	11 x 50	**30.**	11 x 70	**31.**	11 x 90	**32.**	9 x 20
33.	9 x 40	**34.**	9 x 60	**35.**	9 x 80	**36.**	8 x 30
37.	8 x 50	**38.**	8 x 70	**39.**	8 x 90	**40.**	2 x 20
41.	2 x 40	**42.**	2 x 60	**43.**	2 x 80	**44.**	10 x 30
45.	10 x 50	**46.**	10 x 70	**47.**	10 x 90	**48.**	16 x 20
49.	18 x 20	**50.**	14 x 30	**51.**	14 x 50	**52.**	18 x 30
53.	18 x 50	**54.**	16 x 40	**55.**	23 x 20	**56.**	14 x 80
57.	32 x 30	**58.**	41 x 20	**59.**	52 x 50	**60.**	61 x 30
61.	58 x 80	**62.**	63 x 40	**63.**	14 x 20	**64.**	18 x 70
65.	78 x 30	**66.**	81 x 20	**67.**	53 x 30	**68.**	85 x 40
69.	14 x 40	**70.**	16 x 60	**71.**	75 x 50	**72.**	54 x 40
73.	23 x 60	**74.**	43 x 40	**75.**	20 x 60	**76.**	33 x 50
77.	18 x 90	**78.**	77 x 70	**79.**	14 x 60	**80.**	16 x 80
81.	55 x 60	**82.**	65 x 50	**83.**	23 x 40	**84.**	45 x 60
85.	14 x 70	**86.**	20 x 40	**87.**	34 x 70	**88.**	67 x 60
89.	23 x 80	**90.**	14 x 90	**91.**	47 x 80	**92.**	20 x 80

Long multiplication

In long multiplication, there are two methods of working. Use the one you may have learnt already or the one your teacher may wish you to use.

EXAMPLE 1   213 x 23

Starting with the left-hand figure:

```
     213
     x23
    4260 Multiply 213 by 20
     639 Multiply 213 by 3
Answer  4899 Add the columns
```

Steps:
(1) Multiply 213 by 2 (tens)
Write down **'0'** for the 'tens'
Then 2 times 3 = 6, write **6**
2 times 1 = 2, write **2**
2 times 2 = 4, write **4**

(2) Multiply 213 by 3 (units)
3 times 3 = 9, write **9**
3 times 1 = 3, write **3**
3 times 2 = 6, write **6**

Starting with the right-hand figure:

```
     213
     x23
     639 Multiply 213 by 3
    4260 Multiply 213 by 20
Answer  4899 Add the columns
```

Steps:
(1) Multiply 213 by 3 (units)
3 times 3 = 9, write **9**
3 times 1 = 3, write **3**
3 times 2 = 6, write **6**

(2) Multiply 213 by 2 (tens)
Write down **'0'** for the 'tens'
Then 2 times 3 = 6, write **6**
2 times 1 = 2, write **2**
2 times 2 = 4, write **4**

Now add the columns to obtain **4899**.

EXAMPLE 2   326 x 34

```
     326
     x34
    9780 Multiply by 30
    1304 Multiply by 4
Answer  11084 Add
```

```
     326
     x34
    1304 Multiply by 4
    9780 Multiply by 30
Answer  11084 Add
```

Take care with 'carrying figures'.

Exercise 44

1. 21 x 13	**2.** 32 x 13	**3.** 21 x 15	**4.** 23 x 13
5. 122 x 13	**6.** 222 x 14	**7.** 313 x 23	**8.** 324 x 22
9. 121 x 21	**10.** 22 x 14	**11.** 33 x 13	**12.** 232 x 23
13. 212 x 34	**14.** 314 x 21	**15.** 423 x 23	**16.** 241 x 22
17. 32 x 14	**18.** 231 x 33	**19.** 23 x 15	**20.** 432 x 32

21. 25 x 18	**22.** 82 x 14	**23.** 28 x 16	**24.** 34 x 17
25. 52 x 15	**26.** 46 x 13	**27.** 36 x 18	**28.** 42 x 19
29. 35 x 21	**30.** 37 x 22	**31.** 43 x 17	**32.** 47 x 23
33. 39 x 26	**34.** 23 x 17	**35.** 41 x 23	**36.** 34 x 28
37. 46 x 25	**38.** 51 x 21	**39.** 38 x 24	**40.** 54 x 23
41. 24 x 16	**42.** 49 x 22	**43.** 56 x 27	**44.** 63 x 26
45. 68 x 18	**46.** 71 x 29	**47.** 27 x 19	**48.** 39 x 27
49. 29 x 21	**50.** 84 x 25	**51.** 66 x 33	**52.** 87 x 26
53. 131 x 23	**54.** 145 x 27	**55.** 154 x 32	**56.** 168 x 35
57. 173 x 41	**58.** 186 x 44	**59.** 192 x 53	**60.** 197 x 56
61. 209 x 22	**62.** 225 x 25	**63.** 232 x 27	**64.** 243 x 29
65. 314 x 18	**66.** 321 x 24	**67.** 345 x 35	**68.** 362 x 43
69. 402 x 26	**70.** 437 x 32	**71.** 453 x 46	**72.** 481 x 58
73. 524 x 21	**74.** 632 x 34	**75.** 718 x 42	**76.** 845 x 56
77. 926 x 36	**78.** 859 x 67	**79.** 786 x 83	**80.** 827 x 94

Multiplication in algebra

EXAMPLES   2a x 3 means 2a   multiplied by 3 = **6a**
                   3b x 5 means 3b   multiplied by 5 = **15b**
                   5c x 1 means 5c   multiplied by 1 = **5c**
                   d x 4 means   d   multiplied by 4 = **4d**

Exercise 45

Simplify the following:

1. 2a x 2	**2.** 3b x 1	**3.** c x 2	**4.** 3d x 0	**5.** 3a x 2
6. 2b x 3	**7.** 3c x 3	**8.** d x 1	**9.** 5a x 1	**10.** b x 5
11. 2c x 4	**12.** 4d x 2	**13.** 3a x 4	**14.** 4b x 3	**15.** 5c x 1
16. 5d x 3	**17.** 4a x 0	**18.** 5b x 4	**19.** 4c x 4	**20.** 5d x 5
21. 4a x 7	**22.** 7b x 6	**23.** 6c x 8	**24.** 8d x 9	**25.** 7a x 8
26. 6b x 9	**27.** 8c x 6	**28.** 9d x 0	**29.** 12a x 6	**30.** 8b x 8
31. 11c x 11	**32.** 9d x 12	**33.** 10a x 4	**34.** 4b x 10	**35.** 6d x 0
36. 12a x 5	**37.** 5a x 12	**38.** 6b x 8	**39.** 8b x 6	**40.** 9c x 7
41. 7c x 9	**42.** 7d x 8	**43.** 8d x 7	**44.** 6a x 9	**45.** 9a x 6
46. 4b x 7	**47.** 7b x 5	**48.** 5c x 9	**49.** 9c x 8	**50.** 8d x 12

14 How many? – Division

Easy division

EXAMPLES $36 \div 4$ is asking 'how many 4s in 36?' Answer = 9
$42 \div 6$ is asking 'how many 6s in 42?' Answer = 7
$54 \div 9$ is asking 'how many 9s in 54?' Answer = 6

Such questions are easily answered if you know your multiplication tables — you should have learnt them by now.

Exercise 46

1. $18 \div 6$	2. $24 \div 6$	3. $32 \div 8$	4. $45 \div 9$	5. $42 \div 7$
6. $10 \div 10$	7. $7 \div 1$	8. $121 \div 11$	9. $54 \div 6$	10. $60 \div 12$
11. $60 \div 5$	12. $18 \div 9$	13. $36 \div 3$	14. $20 \div 10$	15. $16 \div 4$
16. $84 \div 12$	17. $110 \div 11$	18. $21 \div 7$	19. $28 \div 7$	20. $16 \div 8$
21. $42 \div 6$	22. $66 \div 6$	23. $27 \div 9$	24. $24 \div 8$	25. $88 \div 11$
26. $40 \div 10$	27. $42 \div 7$	28. $72 \div 12$	29. $55 \div 5$	30. $36 \div 9$
31. $20 \div 4$	32. $45 \div 9$	33. $33 \div 3$	34. $60 \div 10$	35. $9 \div 1$
36. $48 \div 12$	37. $132 \div 11$	38. $40 \div 8$	39. $24 \div 4$	40. $72 \div 6$
41. $60 \div 6$	42. $35 \div 7$	43. $63 \div 9$	44. $66 \div 11$	45. $32 \div 8$
46. $80 \div 10$	47. $50 \div 5$	48. $36 \div 12$	49. $30 \div 6$	50. $33 \div 11$
51. $54 \div 9$	52. $8 \div 8$	53. $30 \div 3$	54. $49 \div 7$	55. $56 \div 8$
56. $56 \div 7$	57. $24 \div 12$	58. $100 \div 10$	59. $81 \div 9$	60. $28 \div 4$
61. $45 \div 5$	62. $12 \div 4$	63. $27 \div 3$	64. $32 \div 4$	65. $120 \div 10$
66. $55 \div 11$	67. $72 \div 8$	68. $44 \div 11$	69. $48 \div 8$	70. $24 \div 3$
71. $63 \div 7$	72. $40 \div 5$	73. $48 \div 6$	74. $110 \div 10$	75. $99 \div 9$
76. $88 \div 8$	77. $77 \div 11$	78. $36 \div 4$	79. $108 \div 12$	80. $21 \div 3$
81. $18 \div 6$	82. $18 \div 3$	83. $72 \div 9$	84. $35 \div 5$	85. $77 \div 7$
86. $96 \div 12$	87. $36 \div 6$	88. $15 \div 3$	89. $96 \div 8$	90. $40 \div 4$
91. $90 \div 10$	92. $22 \div 11$	93. $120 \div 12$	94. $108 \div 9$	95. $99 \div 11$
96. $84 \div 7$	97. $12 \div 3$	98. $70 \div 10$	99. $9 \div 3$	100. $30 \div 5$
101. $6 \div 3$	102. $25 \div 5$	103. $24 \div 2$	104. $50 \div 10$	105. $24 \div 6$
106. $14 \div 7$	107. $132 \div 12$	108. $80 \div 8$	109. $44 \div 4$	110. $90 \div 9$

Dividing by 10, 100, 1000

From the multiplication tables we can see such facts as:

$$40 \div 10 \text{ (the number of 10s in 40)} = 4$$
$$70 \div 10 \text{ (the number of 10s in 70)} = 7$$

We just remove the nought from the number we are using:

$$200 \div 10 = \text{(the number of 10s in 200)} = 20$$
$$6000 \div 10 = \text{(the number of 10s in 6000)} = 600$$

When dividing by 10 we remove one nought from the end of the number.

$$200 \div 100 \text{ (the number of 100s in 200)} = 2$$
$$5000 \div 100 \text{ (the number of 100s in 5000)} = 50$$

When dividing by 100 we remove two noughts from the end of the number.

$$5000 \div 1000 \text{ (the number of 1000s in 5000)} = 5$$
$$30\,000 \div 1000 \text{ (the number of 1000s in 30\,000)} = 30$$

When dividing by 1000 we remove three noughts from the end of the number.

Exercise 47

1. $120 \div 10$	2. $7000 \div 100$	3. $600 \div 100$	4. $6800 \div 10$
5. $5000 \div 1000$	6. $90 \div 10$	7. $3500 \div 100$	8. $46\,000 \div 1000$
9. $1000 \div 10$	10. $110 \div 10$	11. $700 \div 100$	12. $10\,000 \div 100$
13. $130 \div 10$	14. $1200 \div 100$	15. $5700 \div 100$	16. $30 \div 10$
17. $7900 \div 100$	18. $3000 \div 1000$	19. $800 \div 10$	20. $9000 \div 100$
21. $70 \div 10$	22. $900 \div 100$	23. $6900 \div 10$	24. $500 \div 10$
25. $4000 \div 1000$	26. $80 \div 10$	27. $100 \div 100$	28. $12\,000 \div 100$
29. $700 \div 10$	30. $350 \div 10$	31. $400 \div 100$	32. $11\,000 \div 1000$
33. $460 \div 10$	34. $20 \div 10$	35. $6000 \div 100$	36. $83\,000 \div 10$
37. $200 \div 100$	38. $1000 \div 100$	39. $240 \div 10$	40. $50 \div 10$
41. $8000 \div 100$	42. $1100 \div 10$	43. $24\,000 \div 1000$	44. $10 \div 10$
45. $2000 \div 10$	46. $300 \div 100$	47. $58\,000 \div 100$	48. $60 \div 10$

Dividing by 20, 30, 40 etc.

EXAMPLE 1. $600 \div 30 = 20$ This is easier in the form of a **fraction**.

STEPS:

$$\frac{60\cancel{0}}{3\cancel{0}}$$

We **cancel** the '0' underneath with the last '0' on top.
(**Dividing by 10**)

$$\frac{20}{\cancel{60}}{\cancel{3}}$$

We now divide the '3' underneath into the '60' on top.
(cancelling)

Ans = 20

EXAMPLE 2.

   $4000 \div 20$

$= \dfrac{400\cancel{0}}{2\cancel{0}}$

$= \dfrac{\overset{200}{\cancel{400}}}{\cancel{2}}$

Ans = 200

EXAMPLE 3.

   $8000 \div 2000$

$= \dfrac{800\cancel{0}\cancel{0}\cancel{0}}{200\cancel{0}\cancel{0}\cancel{0}}$

$= \dfrac{\overset{4}{\cancel{8}}}{\cancel{2}}$

Ans = 4

Exercise 48

1. $40 \div 20$	**2.** $60 \div 30$	**3.** $80 \div 40$	**4.** $100 \div 50$
5. $200 \div 40$	**6.** $300 \div 60$	**7.** $400 \div 80$	**8.** $500 \div 50$
9. $600 \div 20$	**10.** $800 \div 40$	**11.** $900 \div 90$	**12.** $1000 \div 20$
13. $1000 \div 50$	**14.** $2000 \div 40$	**15.** $3000 \div 60$	**16.** $4000 \div 80$
17. $150 \div 30$	**18.** $160 \div 40$	**19.** $180 \div 20$	**20.** $140 \div 70$
21. $120 \div 60$	**22.** $250 \div 50$	**23.** $360 \div 90$	**24.** $240 \div 80$
25. $800 \div 200$	**26.** $600 \div 300$	**27.** $550 \div 50$	**28.** $900 \div 300$
29. $3000 \div 300$	**30.** $3000 \div 30$	**31.** $5000 \div 50$	**32.** $5000 \div 500$
33. $4000 \div 200$	**34.** $8000 \div 400$	**35.** $6000 \div 300$	**36.** $7000 \div 70$
37. $300 \div 50$	**38.** $600 \div 120$	**39.** $700 \div 50$	**40.** $600 \div 50$
41. $1200 \div 60$	**42.** $1200 \div 400$	**43.** $1200 \div 30$	**44.** $1200 \div 600$
45. $2400 \div 800$	**46.** $3600 \div 600$	**47.** $4800 \div 400$	**48.** $6000 \div 500$
49. $7200 \div 1200$	**50.** $8400 \div 700$	**51.** $9600 \div 800$	**52.** $8100 \div 90$

Easy division in algebra

EXAMPLES

   $4x \div 2 = 2x$ $7y \div 1 = 7y$

   $9a \div 3 = 3a$ $5c \div 5 = c$

Exercise 49

1. $4a \div 2$	**2.** $6b \div 2$	**3.** $8c \div 2$	**4.** $6x \div 2$
5. $10x \div 5$	**6.** $3d \div 1$	**7.** $5a \div 5$	**8.** $10a \div 2$
9. $8b \div 4$	**10.** $12x \div 2$	**11.** $16c \div 8$	**12.** $12a \div 4$
13. $16d \div 4$	**14.** $18x \div 2$	**15.** $7b \div 7$	**16.** $11c \div 1$
17. $18a \div 9$	**18.** $12d \div 6$	**19.** $16b \div 2$	**20.** $24d \div 24$
21. $24a \div 2$	**22.** $36x \div 12$	**23.** $48c \div 3$	**24.** $24b \div 6$
25. $36x \div 6$	**26.** $48d \div 6$	**27.** $48a \div 8$	**28.** $24x \div 12$
29. $36c \div 18$	**30.** $32b \div 16$	**31.** $105x \div 5$	**32.** $45a \div 9$

Short division

EXAMPLE 1. $126 \div 6$

$$
\begin{array}{r}
21 \\
6\,)\overline{126}
\end{array}
$$

Ans = 21

STEPS:
6 into 12 goes 2 (exactly); write **2**
6 into 6 goes 1 (exactly); write **1**

EXAMPLE 2. $288 \div 8$

$$
\begin{array}{r}
3\;6 \\
8\,)\,2\;8^{4}8
\end{array}
$$

Ans = 36

STEPS:
8 into 28 goes 3; (24 and 4 over); write **3**
Put the 4 up to the 8 to give 48
8 into 48 goes 6 (exactly); write **6**

Exercise 50

1. $42 \div 3$	2. $56 \div 4$	3. $75 \div 5$	4. $84 \div 6$
5. $98 \div 7$	6. $104 \div 8$	7. $117 \div 9$	8. $143 \div 11$
9. $168 \div 12$	10. $54 \div 3$	11. $64 \div 4$	12. $85 \div 5$
13. $96 \div 6$	14. $105 \div 7$	15. $144 \div 8$	16. $198 \div 9$
17. $187 \div 11$	18. $192 \div 12$	19. $72 \div 3$	20. $76 \div 4$
21. $138 \div 6$	22. $133 \div 7$	23. $192 \div 8$	24. $216 \div 9$
25. $90 \div 5$	26. $253 \div 11$	27. $180 \div 12$	28. $87 \div 3$
29. $203 \div 7$	30. $216 \div 8$	31. $243 \div 9$	32. $252 \div 12$
33. $111 \div 3$	34. $92 \div 4$	35. $115 \div 5$	36. $204 \div 6$
37. $256 \div 8$	38. $252 \div 9$	39. $374 \div 11$	40. $384 \div 12$
41. $348 \div 12$	42. $255 \div 3$	43. $136 \div 4$	44. $230 \div 5$
45. $342 \div 6$	46. $336 \div 7$	47. $392 \div 8$	48. $378 \div 9$
49. $495 \div 11$	50. $636 \div 12$	51. $282 \div 3$	52. $356 \div 4$
53. $365 \div 5$	54. $414 \div 6$	55. $504 \div 7$	56. $624 \div 8$
57. $567 \div 9$	58. $671 \div 11$	59. $936 \div 12$	60. $462 \div 3$

Long division

You will probably need your teacher's help with these examples:

EXAMPLE 1. $575 \div 23$ or $\dfrac{575}{23}$

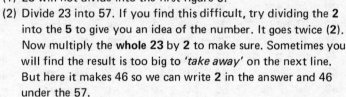

STEPS:

(1) 23 will not divide into the first figure 5.

(2) Divide 23 into 57. If you find this difficult, try dividing the **2** into the **5** to give you an idea of the number. It goes twice (**2**). Now multiply the **whole 23** by 2 to make sure. Sometimes you will find the result is too big to *'take away'* on the next line. But here it makes 46 so we can write 2 in the answer and 46 under the 57.

(3) Taking 46 away from 57 leaves 11. Now bring down the next figure, 5.

(4) Divide 23 into 115. If you find this difficult, try dividing the **2** into the **11** to give you an idea of the number. It goes 5 times. Now multiply the **whole 23** by 5 to make sure. It makes 115 which is just right so we can write 5 in the answer and 115 under the 115 to take away. Nothing is left so the division is finished.

NOTE If the result had been larger than 115, it would have been *'too big to take away'*. In this case, the answer figure would have to be smaller than 5. Perhaps 4 or even 3.

EXAMPLE 2. $7303 \div 67$ or $\dfrac{7303}{67}$

```
          *
        109
   67 ) 7303
        67↓↓
  A     60 |
        00 ↲
  B     603
        603
        · · ·

  Ans = 109
```

STEPS:

(1) Divide 67 into 73, it goes once so write 1 in the answer line.

(2) Multiply 67 by 1 to give 67 which is now taken away from 73 to leave 6. Bring down the next figure, 0.

(3) At stage A, we are dividing 67 into 60 and *'it won't go'* therefore the digit **'0'** must appear in the answer (*). This is very important to show that it goes '0' times.

(4) Multiply 67 by 0 to give 00 which is now taken away from 60 to leave 60. Bring down the next figure, 3.

(5) At stage B, we are dividing 67 into 603. It goes 9 times. (6 into 60 gives 10 which is too big). Write **9** in the answer.

(6) Last step, take 603 away from 603. Nothing left — the division is finished.

Exercise 51

1. $39 \div 13$	**2.** $42 \div 14$	**3.** $45 \div 15$	**4.** $48 \div 16$
5. $52 \div 13$	**6.** $56 \div 14$	**7.** $75 \div 15$	**8.** $64 \div 16$
9. $78 \div 13$	**10.** $84 \div 14$	**11.** $60 \div 15$	**12.** $80 \div 16$
13. $91 \div 13$	**14.** $98 \div 14$	**15.** $90 \div 15$	**16.** $96 \div 16$
17. $104 \div 13$	**18.** $112 \div 14$	**19.** $120 \div 15$	**20.** $128 \div 16$
21. $34 \div 17$	**22.** $54 \div 18$	**23.** $38 \div 19$	**24.** $63 \div 21$
25. $51 \div 17$	**26.** $72 \div 18$	**27.** $57 \div 19$	**28.** $84 \div 21$
29. $68 \div 17$	**30.** $90 \div 18$	**31.** $76 \div 19$	**32.** $105 \div 21$
33. $85 \div 17$	**34.** $108 \div 18$	**35.** $95 \div 19$	**36.** $126 \div 21$
37. $102 \div 17$	**38.** $126 \div 18$	**39.** $114 \div 19$	**40.** $147 \div 21$
41. $69 \div 23$	**42.** $120 \div 24$	**43.** $150 \div 25$	**44.** $182 \div 26$
45. $115 \div 23$	**46.** $72 \div 24$	**47.** $125 \div 25$	**48.** $156 \div 26$
49. $161 \div 23$	**50.** $144 \div 24$	**51.** $200 \div 25$	**52.** $130 \div 26$
53. $108 \div 27$	**54.** $162 \div 27$	**55.** $140 \div 28$	**56.** $196 \div 28$
57. $186 \div 31$	**58.** $165 \div 33$	**59.** $140 \div 35$	**60.** $111 \div 37$
61. $294 \div 42$	**62.** $264 \div 44$	**63.** $230 \div 46$	**64.** $192 \div 48$
65. $255 \div 51$	**66.** $212 \div 53$	**67.** $165 \div 55$	**68.** $114 \div 57$
69. $116 \div 58$	**70.** $224 \div 56$	**71.** $324 \div 54$	**72.** $416 \div 52$
73. $201 \div 67$	**74.** $325 \div 65$	**75.** $441 \div 63$	**76.** $549 \div 61$
77. $648 \div 72$	**78.** $592 \div 74$	**79.** $456 \div 76$	**80.** $390 \div 78$
81. $648 \div 81$	**82.** $498 \div 83$	**83.** $340 \div 85$	**84.** $261 \div 87$
85. $623 \div 89$	**86.** $810 \div 90$	**87.** $368 \div 92$	**88.** $480 \div 96$
89. $182 \div 14$	**90.** $272 \div 16$	**91.** $288 \div 18$	**92.** $294 \div 21$
93. $437 \div 23$	**94.** $375 \div 25$	**95.** $486 \div 27$	**96.** $609 \div 29$
97. $800 \div 32$	**98.** $782 \div 34$	**99.** $684 \div 36$	**100.** $608 \div 38$
101. $984 \div 41$	**102.** $946 \div 43$	**103.** $765 \div 45$	**104.** $658 \div 47$
105. $686 \div 49$	**106.** $936 \div 52$	**107.** $1242 \div 54$	**108.** $1456 \div 56$
109. $1824 \div 57$	**110.** $2065 \div 59$	**111.** $2394 \div 63$	**112.** $2665 \div 65$
113. $2412 \div 67$	**114.** $3096 \div 72$	**115.** $4104 \div 76$	**116.** $4836 \div 78$
117. $3696 \div 84$	**118.** $3219 \div 87$	**119.** $4823 \div 91$	**120.** $7125 \div 95$
121. $2828 \div 28$	**122.** $3531 \div 33$	**123.** $4368 \div 42$	**124.** $5830 \div 55$
125. $2178 \div 121$	**126.** $3744 \div 234$	**127.** $4424 \div 316$	**128.** $5124 \div 427$
129. $7308 \div 36$	**130.** $9270 \div 45$	**131.** $11\,628 \div 57$	
132. $12\,627 \div 61$	**133.** $16\,006 \div 53$	**134.** $19\,520 \div 64$	
135. $22\,248 \div 72$	**136.** $34\,486 \div 86$	**137.** $4107 \div 37$	
138. $5043 \div 41$	**139.** $7144 \div 47$	**140.** $8823 \div 51$	
141. $7303 \div 67$	**142.** $10\,368 \div 72$	**143.** $12\,879 \div 81$	
144. $15\,660 \div 87$			

5 How long? – Measuring length

The diagram shows the first 10 centimetres of a ruler. AG is a straight line alongside the ruler and it has been marked into smaller pieces shown by the letters (AB, AC, BC etc.); **cm is the abbreviation (short form) for centimetre.** You are measuring in centimetres.

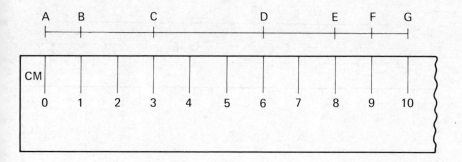

Exercise 52

1. How long are the lines AB, AC, AD, AE, AF, AG?
2. How long are the lines BC, BD, BE, BF, BG?
3. How long are the lines CD, CE, CF, CG?
4. How long are the lines DE, DF, DG?
5. How long are the lines EF, EG?
6. How long is the line FG?
7. How much longer is AG than AF?
8. How much longer is AG than AE?
9. How much longer is AG than AD?
10. How much longer is AG than AC?
11. How much longer is AG than AB?
12. How much longer is AF than AE?
13. How much longer is AF than AD?
14. How much longer is AF than AC?
15. How much longer is AF than AB?
16. How much must be added to AB to make AE?

More measuring — Length

The next diagram shows the first 10 cm of a ruler divided into centimetres and half-centimetres. AG is a straight line alongside the ruler and it has been marked into smaller pieces. PT is a second straight line but this time the measurement of the pieces cannot be made *exactly*, you must judge which mark on the ruler is nearest to the point you are measuring and you give this as your answer. For example, PQ is only just over 2½ cm, it isn't near enough to the next mark on the ruler to be counted as 3 cm, so **2½ cm** would be the *approximate length of PQ*. You have to decide whether or not you are 'more than half-way' to the next mark on the ruler when you are *making approximations*. Remember you are measuring in centimetres so you must write the units in your answers; the abbreviation is **cm** in each case.

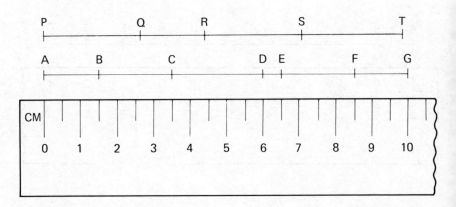

Exercise 53

1. How long are the lines AB, AC, AD, AE, AF, AG?
2. How long are the lines BC, BD, BE, BF, BG?
3. How long are the lines CD, CE, CF, CG?
4. How long are the lines DE, DF, DG?
5. How long are the lines EF, EG?
6. How long is the line FG?
7. How much longer is AG than AF?
8. How much longer is AG than AE?
9. How much longer is AG than AD?
10. How much longer is AG than AC?
11. How much longer is AG than AB?
12. How many ½ cm are there in 1 cm, 2 cm, 3 cm, 4 cm, 5 cm, 6 cm, 7 cm, 8 cm, 9 cm, 10 cm?
13. How many ½ cm are there in 1½ cm, 2½ cm, 3½ cm, 4½ cm, 5½ cm, 6½ cm, 7½ cm, 8½ cm, 9½ cm, 10½ cm, ½ cm?

14. How much longer is AF than AE?
15. How much longer is AF than AD?
16. How much longer is AF than AC?
17. How much longer is AF than AB?
18. How much longer is AE than AD?
19. How much longer is AE than AC?
20. How much longer is AE than AB?
21. How much must be added to AB to make AD?
22. How much must be added to AB to make AC?
23. How much must be added to AC to make AD?
24. How much must be added to BC to make BD?
25. What is the approximate length of PQ, PR, PS, PT?

NOTE: You may need your ruler to help you with the rest of these questions.

26. What is the approximate length of QR, QS, QT?
27. What is the approximate length of RS, RT?
28. What is the approximate length of ST?

Decimetres, centimetres, millimetres

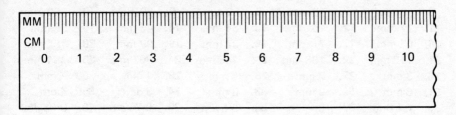

Once more we see the first 10 cm of a ruler and we can learn two new words. A length of 10 cm is called a **decimetre** (abbreviation **dm**). Each centimetre is divided into very tiny parts called **millimetres** (abbreviation **mm**).

Questions:
(1) How many millimetres are there in each centimetre?
(2) How many millimetres are there in each ½ cm?
(3) How many millimetres are there in one decimetre?

Learn this table
        10 millimetres (mm) = 1 centimetre (cm)
        10 centimetres (cm) = 1 decimetre (dm)
        100 millimetres (mm) = 1 decimetre (dm)

Lengths may be written in several ways

EXAMPLES   *Decimal points are used to separate the units and fractions of units.*

$$
\begin{aligned}
5 \text{ mm} &= 0{\cdot}5 \text{ cm} = 0{\cdot}05 \text{ dm} \\
8 \text{ mm} &= 0{\cdot}8 \text{ cm} = 0{\cdot}08 \text{ dm} \\
16 \text{ mm} &= 1{\cdot}6 \text{ cm} = 0{\cdot}16 \text{ dm} \\
19 \text{ mm} &= 1{\cdot}9 \text{ cm} = 0{\cdot}19 \text{ dm} \\
24 \text{ mm} &= 2{\cdot}4 \text{ cm} = 0{\cdot}24 \text{ dm} \\
136 \text{ mm} &= 13{\cdot}6 \text{ cm} = 1{\cdot}36 \text{ dm}
\end{aligned}
$$

In fact we don't use ½ cm, instead we say 0·5 cm or 5 mm.

NOTE **Each time we divide by 10, the decimal point moves one place to the left. When we multiply by 10, the point moves one place to the right. When multiplying or dividing by 100, the decimal point moves two places.**

Exercise 54

Turn the following into centimetres (cm):

1.	10 mm	**2.**	20 mm	**3.**	30 mm	**4.**	40 mm	**5.**	50 mm
6.	60 mm	**7.**	70 mm	**8.**	80 mm	**9.**	90 mm	**10.**	100 mm
11.	12 mm	**12.**	15 mm	**13.**	34 mm	**14.**	46 mm	**15.**	58 mm
16.	63 mm	**17.**	71 mm	**18.**	87 mm	**19.**	99 mm	**20.**	104 mm
21.	112 mm	**22.**	136 mm	**23.**	245 mm	**24.**	367 mm	**25.**	484 mm
26.	3 mm	**27.**	2 mm	**28.**	5 mm	**29.**	4 mm	**30.**	7 mm
31.	6 mm	**32.**	9 mm	**33.**	8 mm	**34.**	3 dm	**35.**	4 dm
36.	6·5 dm	**37.**	8·2 dm	**38.**	0·8 dm	**39.**	0·75 dm	**40.**	0·06 dm

Exercise 55

Turn the following into millimetres (mm):

1.	1 cm	**2.**	2 cm	**3.**	3 cm	**4.**	4 cm	**5.**	5 cm
6.	6 cm	**7.**	7 cm	**8.**	8 cm	**9.**	9 cm	**10.**	10 cm
11.	1 dm	**12.**	1·3 cm	**13.**	2·5 cm	**14.**	3·7 cm	**15.**	4·9 cm
16.	5·2 cm	**17.**	6·4 cm	**18.**	8·6 cm	**19.**	10·8 cm	**20.**	20 cm
21.	21·3 cm	**22.**	33·5 cm	**23.**	45·7 cm	**24.**	57·9 cm	**25.**	63 cm
26.	3 dm	**27.**	5 dm	**28.**	7 dm	**29.**	9 dm	**30.**	1·2 dm
31.	2·4 dm	**32.**	3·6 dm	**33.**	4·8 dm	**34.**	5 dm	**35.**	6·1 dm
36.	6·3 dm	**37.**	7·5 dm	**38.**	0·6 dm	**39.**	0·82 dm	**40.**	0·05 dm

Exercise 56

Turn the following into decimetres (dm):

1. 10 cm	**2.** 20 cm	**3.** 30 cm	**4.** 40 cm	**5.** 50 cm
6. 600 mm	**7.** 700 mm	**8.** 800 mm	**9.** 900 mm	**10.** 1000 mm
11. 15 cm	**12.** 25 cm	**13.** 34 cm	**14.** 46 cm	**15.** 58 cm
16. 63 cm	**17.** 79 cm	**18.** 82 cm	**19.** 98 cm	**20.** 106 cm
21. 214 cm	**22.** 323 cm	**23.** 436 cm	**24.** 561 cm	**25.** 679 cm
26. 121 mm	**27.** 352 mm	**28.** 573 mm	**29.** 734 mm	**30.** 916 mm
31. 13·6 cm	**32.** 17·5 cm	**33.** 21·4 cm	**34.** 29·2 cm	**35.** 34·3 cm
36. 38·2 cm	**37.** 42·7 cm	**38.** 57·9 cm	**39.** 71·8 cm	**40.** 92·5 cm

NOTE You may need your ruler to help you with the next exercise.

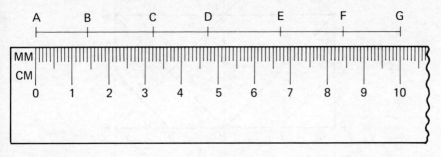

Exercise 57

1. Measure these lines in millimetres: AB, AC, AD, AE, AF, AG.
2. Measure these lines in millimetres: BC, BD, BE, BF, BG.
3. Measure these lines in centimetres: CD, CE, CF, CG.
4. Measure these lines in centimetres: DE, DF, DG.
5. Measure these lines in decimetres: EF, EG.
6. Measure these lines in decimetres: FG, AG.
7. How much longer is AG than AF (in cm)?
8. How much longer is AG than AE (in cm)?
9. How much longer is AG than AD (in cm)?
10. How much longer is AG than AC (in cm)?
11. How much longer is AG than AB (in cm)?
12. How much longer is AF than AE (in mm)?
13. How much longer is AF than AD (in mm)?
14. How much longer is AF than AC (in mm)?
15. How much longer is AF than AB (in mm)?
16. How much longer is AE than AD (in cm)?
17. How much longer is AE than AC (in mm)?
18. How much longer is AE than AB (in dm)?
19. How much longer is AD than AC (in dm)?
20. How much longer is AD than AB (in dm)?

Exercise 58

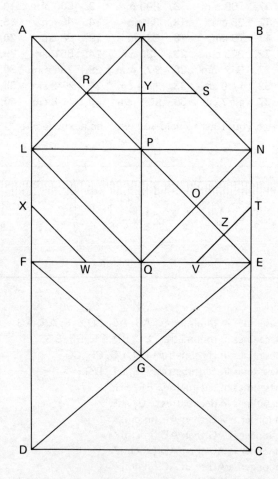

1. Use the diagram above to measure the lengths of the lines listed in the following table. Copy the table and complete the entries.

LINE	LENGTH IN MILLIMETRES	LENGTH IN CENTIMETRES
AL		
AX		
AF		
AD		
AM		
AB		
AP		
AE		
FG		
FO		
LQ		
XW		
RS		
RY		
OE		
OZ		
MY		
MG		
QG		
WV		

2. Measure the lengths of the following lines, in millimetres, and enter the results in a table like the one given overleaf:

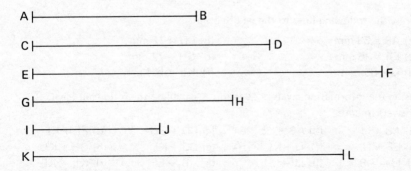

LINE	LENGTH IN MILLIMETRES
AB	
CD	
EF	
GH	
IJ	
KL	

Use your results to calculate the following:

(a) AB + CD   (b) AB + GH   (c) AB + IJ   (d) GH + IJ
(e) AB + EF   (f) EF − KL   (g) KL − IJ   (h) EF − GH

3. Measure the lengths of the lines below, in cm, and put the results in a table:

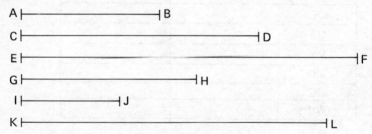

Use your results to calculate the following:

(a) CD + GH   (b) IJ + KL   (c) AB + EF   (d) EF + IJ
(e) AB + IJ   (f) AB + KL   (g) CD + IJ   (h) EF + KL
(i) CD − IJ   (j) EF − KL   (k) EF − AB   (l) KL − GH

4. Draw the following lines to the length given:

(a) AB = 53 mm   (b) CD = 77 mm
(c) EF = 36 mm   (d) GH = 72 mm
(e) IJ = 89 mm   (f) KL = 104 mm

5. Using the information given in question 4, calculate the following (give answers in cm):

(a) AB + IJ   (b) AB + KL   (c) CD + GH   (d) EF + KL
(e) EF + IJ   (f) IJ + KL   (g) CD + KL   (h) GH + KL
(i) IJ − AB   (j) IJ − CD   (k) AB − EF   (l) KL − AB

5 Measuring in algebra

EXAMPLES: If AB = x mm, BC = $2x$ mm and CD = $3x$ mm, then AD = $6x$ mm.

If AD = 9 cm and AB = x cm, BD = $(9 - x)$ cm.

NOTE The brackets are used to show that the whole thing is in cm.

Exercise 59

1. If AB = a cm and BD = b cm, how long is AD?
2. If BC = p mm and CD = q mm, how long is BD?
3. If AB = x cm, BC = y cm and CD = z cm, how long is AD?
4. If AB = a mm, BC = b mm and CD = c mm, how long is AD?
5. If AB = x cm, BC = x cm and CD = x cm, how long is AD?
6. If AB = x mm, BC = $2x$ mm and CD = $3x$ mm, how long is AD?
7. If AD = 6 cm and AC = 4 cm, how long is CD?
8. If AD = 6 cm and AC = x cm, how long is CD?
9. If AD = x cm and AC = y cm, how long is CD?
10. If AD = a mm and AB = b mm, how long is BD?
11. If AD = $6x$ cm and AC = $4x$ cm, how long is CD?
12. If AD = $8x$ cm, AB = $2x$ cm and BC = $3x$ cm, how long is CD?

17 Measuring angles

When we measure angles we are measuring the **amount of turning** which takes place. If we make a complete turn through a full circle it is described as *turning through 360 degrees* — written as **360°**. Each degree can be divided into smaller parts but we shall not worry about these for the present.

Consider the **hour hand** of a clock starting at 12 o'clock:

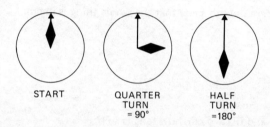

START   QUARTER   HALF
             TURN   TURN
            = 90°   =180°

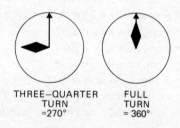

THREE—QUARTER   FULL
    TURN   TURN
   =270°   = 360°

The right angle

Of the many angles which we shall discuss, the **right angle** is the most important. When we look at a post in the ground, we judge whether it is **upright** by the angle it makes with the ground.

The corners of the pages of this book are right angles, they are all quarters of full circles, that is they are each angles of **90 degrees (90°)**.

Angles which are less than 90° are called acute angles; angles greater than 90° are called obtuse angles.

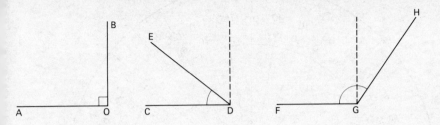

Angle AOB is a **right angle,** note how we show this fact by marking the angle ⌐.

Angle CDE is less than 90°, it is an **acute angle.** Note how we are using three letters to indicate the angles, the middle letter gives the position of the angle; that is, where the two arms meet. We can sometimes use one letter, in this case angle D, but it can be confusing as you will see when you come to exercise 59.

Angle FGH (or angle G) is greater than a right angle; it is an **obtuse angle.** There is a *symbol for the word angle*; it is ∠. So we can say we have been discussing:

> angle AOB or ∠AOB or ∠O
> angle CDE or ∠CDE or ∠C
> angle FGH or ∠FGH or ∠G

Questions:
1. Name some parts of the room, or objects in the room, where right angles are to be found.
2. Which of the following are:
    (a) right angles (b) acute angles (c) obtuse angles?

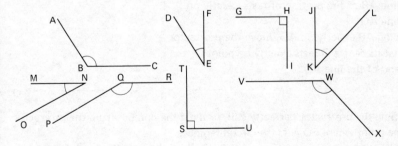

3. Does the *length* of the arms of an angle affect the size of the angle between the arms?

73

The protractor

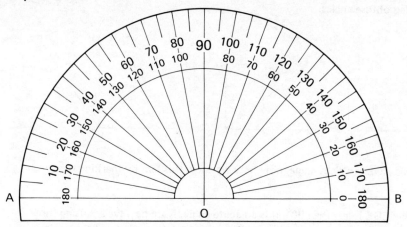

Angles are measured with the aid of **protractors**.

If you examine your protractor you will notice, along the straight edge, a narrow band of material without any graduation marks. This strip is similar to the waste material at each end of a ruler, it is there to prevent the important part becoming damaged.

Angles are measured from the line AO or the line BO in the diagram.

The next important thing to notice is the **centre** of the protractor, this is the point O in the diagram, where the line showing 90° meets the base line AB.

Next, you should see that there are two sets of figures round the circumference of the protractor. To make sure you are using the correct set of figures, you must first decide whether the angle you are measuring is more or less than a right angle. That is, *is your angle more or less than 90°?*

Let us measure ∠AOF:

1. Is ∠AOF more or less than a right angle?
   (Answer = Less)
2. Place your protractor over the drawing so that the line AB of the protractor rests exactly on the line AO.
3. Keeping AB exactly on AO, move the protractor so that its centre O rests exactly on point O, the end of the line AO.

4. Keeping the protractor perfectly still, read off the number from the protractor at the point where FO cuts the circumference.
5. There are two numbers, 40° and 140°. Which of these shall we give for the answer? We decided at the beginning that ∠AOF was less than a right angle, that is less than 90°. **Therefore the answer must be 40°.**

Let us measure ∠BOH:

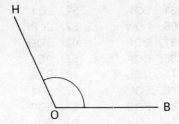

1. Is LBOH **acute** or **obtuse**?
2. Place your protractor over the drawing so that the line AB of the protractor rests exactly on the line OB.
3. Keeping AB exactly on OB, move the protractor so that its centre O rests exactly on point O, the end of the line OB.
4. Keeping the protractor perfectly still, read off the number from the protractor at the point where HO cuts the circumference.
5. There are two numbers, 65° and 115°. Which of these shall we give for the answer?

Exercise 60

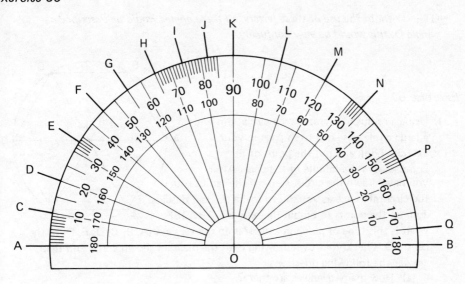

Read off the values of the following angles:

1. AOC	2. AOD	3. AOE	4. AOF	5. AOG
6. AOH	7. AOI	8. AOJ	9. AOK	10. AOL
11. AOM	12. AON	13. AOP	14. AOQ	15. AOB
16. BOQ	17. BOP	18. BON	19. BOM	20. BOL
21. BOK	22. BOJ	23. BOI	24. BOH	25. BOG
26. BOF	27. BOE	28. BOD	29. BOC	30. BOA
31. KOJ	32. KOI	33. KOH	34. KOG	35. KOF
36. KOE	37. KOD	38. KOC	39. KOA	40. KOL
41. KOM	42. KON	43. KOP	44. KOQ	45. KOB
46. COD	47. COE	48. COF	49. COG	50. COH

51. COI	52. COJ	53. COK	54. COL	55. COM
56. CON	57. COP	58. COQ	59. COB	60. DOE
61. DOF	62. DOG	63. DOH	64. DOI	65. DOJ
66. DOK	67. DOL	68. DOM	69. DON	70. DOP
71. DOQ	72. DOB	73. EOF	74. EOG	75. EOH
76. EOI	77. EOJ	78. EOK	79. EOL	80. EOM
81. EON	82. EOP	83. EOQ	84. EOB	85. FOG
86. FOH	87. FOI	88. FOJ	89. FOK	90. FOL
91. FOM	92. FON	93. FOP	94. FOQ	95. FOB
96. GOH	97. GOI	98. GOJ	99. GOK	100. GOL
101. GOM	102. GON	103. GOP	104. GOQ	105. GOB
106. HOI	107. HOJ	108. HOK	109. HOL	110. HOM
111. HON	112. HOP	113. HOQ	114. HOB	115. IOJ
116. IOK	117. IOL	118. IOM	119. ION	120. IOQ

NOTE *Without the use of* **three** *letters, all these angles would be described as angle O; this would be very confusing.*

Exercise 61

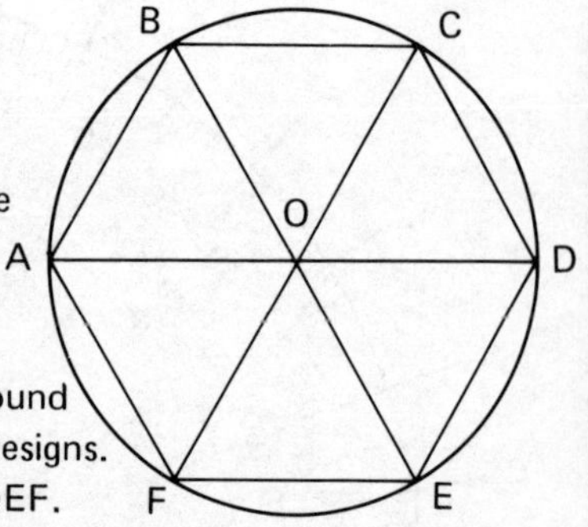

1. Using a radius of 6 cm, draw a circle; this will be a large circle so take a fresh page in your exercise book and put the point of the compasses in the middle of the page. After you have drawn the circle, step off the radius round the circumference as you did to make the leaf designs. Join up the points to obtain the polygon ABCDEF. What is this type of polygon called? Join the vertices (A, B, C, D, E, F) to the centre O. Check your own drawing with the diagram shown above and then answer the following questions.

   (a) How many triangles are there?

   (b) How many angles are there in each triangle?

   (c) Measure every angle in every triangle. What do you notice about the results?

   (d) What should be the size of ∠ABC? Measure it to check your answer.

   (e) What can you say about the following angles ∠BCD, ∠CDE, ∠DEF, ∠EFA, ∠FAB?

   (f) What should be the size of ∠AOC? Measure it to check your answer.

   (g) What is the size of ∠AOD? This is sometimes called a **straight angle**. Can you suggest why?

   (h) What is the sum (total) of all the angles at O, the centre of the circle? ∠AOB + ∠BOC + ∠COD + ∠DOE + ∠EOF + ∠FOA = ?

   (i) Name two other straight angles.

2. Repeat the construction of question 1 but join alternate points to obtain triangle ABC as shown in the diagram. Join the vertices A, B and C to the centre O. Letter your diagram and answer the following questions.

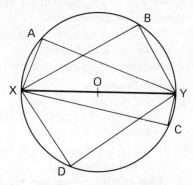

(a) Measure each angle of triangle ABC. What can you say about the results?

(b) Measure each of the angles at O, the centre of the circle (i.e. ∠AOB, ∠BOC, ∠COA). What can you say about the results?

(c) What is the sum of the angles at O, the centre of the circle? ∠AOB + ∠BOC + ∠COA =   ?

(d) Measure ∠OAB and ∠OAC. What can you say about the results?

(e) What does the line OA do to ∠BAC?

(f) The word **'bisect'** means **'cut into two equal parts'**. Does the line OB bisect ∠ABC? Check by measuring ∠ABO and ∠CBO.

(g) Does the line OC bisect ∠ACB? Check by measuring ∠BCO and ∠ACO.

(h) What is the sum of the angles of triangle ABC? ∠ABC + ∠BCA + ∠CAB =   ?

3. Again using a radius of 6 cm, draw a circle. Through O, the centre of the circle, draw a straight line XY cutting the circumference at points X and Y. **XY is called a diameter of the circle.** Select four points A, B, C and D anywhere on the circumference, A and B on one side of the diameter XY, C and D on the other side. Join these four points to X and Y and letter your drawing as shown in the diagram. Answer the following questions.

(a) Look back at the diagram for question 1. How many diameters are shown? Name them.

(b) Does the diagram for question 2 show any diameters? What name do we give to lines like AO, BO and CO?

(c) In the diagram for question 3, what is the size of angle XOY? What name may be given to such an angle and why?

(d) XY cuts the circle in half. What is the name given to a half-circle?

(e) Measure the following angles: ∠XAY, ∠XBY, ∠XDY, ∠XCY. What can be said about these angles?

4. With a radius of 6 cm, draw a circle, centre O. Draw a diameter XOY and with the aid of your protractor construct another diameter AOB at right angles (90°) to XOY. Join the points X, A, Y, B and letter your diagram as shown. Answer the following questions.

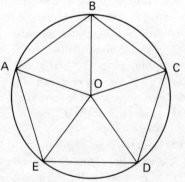

(a) What is the size of ∠XOY and ∠AOB?

(b) If ∠XOA is 90°, what is the size of ∠YOA?

(c) If ∠YOA is 90°, what is the size of ∠YOB?

(d) If ∠YOB is 90°, what is the size of ∠XOB?

(e) What is the sum of the angles at O, the centre of the circle?
   ∠XOA + ∠YOA + ∠YOB + ∠XOB =

(f) Does OA bisect ∠XAY? Check by measuring ∠XAO and ∠YAO.

(g) Does OY bisect ∠AYB? Check by measuring ∠AYO and ∠BYO.

(h) Does OB bisect ∠YBX? Check by measuring ∠YBO and ∠XBO.

(i) Does OX bisect ∠BXA? Check by measuring ∠BXO and ∠AXO.

(j) Measure ∠XAY, ∠AYB, ∠YBX and ∠BXA. What can be said about these angles?

(k) Measure AY, YB, BX and XA. What can be said about these sides?

(l) **A square is a quadrilateral (4 sides) in which all sides are equal in length and all four angles are right angles.** Is the figure AYBX a square?

(m) **AB is called a diagonal of the square AYBX.** Measure AO and BO. What can be said about them?

(n) **XY is called a diagonal of the square AYBX.** Measure XO and YO. What can be said about them?

(o) What do the diagonals of a square do to each other?

5. From this diagram:

(a) What is the sum of the angles at O?
   ∠AOB + ∠BOC + ∠COD + ∠DOE + ∠EOA =   ?

(b) If these five angles are equal, what is the size of each one.

(c) If the points on the circumference are joined, a polygon is obtained. What is the name of the polygon?

(d) Using a circle of radius 6 cm and making each of the angles at O equal to 72°, construct the diagram as shown. Join the vertices, A, B, C, D and E.

(e) Measure ∠ABC, ∠BCD, ∠CDE, ∠DEA, ∠EAB. What can you say about these angles?

(f) What name is given to AO, BO, CO, DO, EO?

6. AOB is a straight line with line CO
cutting AB at O. This forms two
angles ∠AOC and ∠BOC on the
straight line AOB. Angles such as
these are described as **adjacent
angles on a straight line.** Meaning
they are *'next to each other'*.

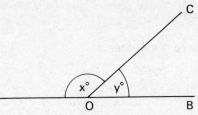

(a) If AOB is a straight line, what can we say about the sum of $x°$ and $y°$?
(b) What can we say about *'the sum of adjacent angles on a straight line'*?
(c) If $x = 120°$, what is the value of y?
(d) If $x = 60°$, what is the value of y?
(e) If $y = 30°$, what is the value of x?
(f) If $y = 140°$, what is the value of x?
(g) If $x = 50°$, what is the value of y?
(h) If $y = 20°$, what is the value of x?
(i) If $x = 70°$, what is the value of y?
(j) If $y = 80°$, what is the value of x?
(k) If x is equal to y, what is the value of each?
(l) If x is twice the size of y, what is the value of each?

7. AOB is a straight line with CO and
DO forming three angles, ∠AOC,
∠COD and ∠BOD.

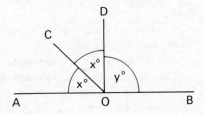

(a) What can we say about the
sum (total) of ∠AOC, ∠COD
and ∠BOD?
(b) If ∠AOC = $x°$ and ∠COD = $x°$
what can we say about ∠AOC
and ∠COD?
(c) If $x = 40°$, what is the value of y?
(d) If $y = 80°$, what is the value of x?
(e) If $y = 90°$, what is the value of x?
(f) If $x = 25°$, what is the value of y?
(g) If $x = 60°$, what is the value of y?
(h) If $y = 30°$, what is the value of x?

Geometry

Things to remember

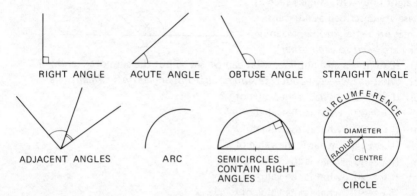

RIGHT ANGLE   ACUTE ANGLE   OBTUSE ANGLE   STRAIGHT ANGLE

ADJACENT ANGLES   ARC   SEMICIRCLES CONTAIN RIGHT ANGLES   CIRCUMFERENCE DIAMETER RADIUS CENTRE CIRCLE

An angle is the amount of turning which has taken place.

Adjacent angles are angles which are next to each other.

Adjacent angles on a straight line add up to 180°.

One complete revolution gives an angle of 360°.

Angles which meet at a point, as they do at the centre of a circle, add up to 360°.

A square is a quadrilateral (4 sides) in which all the sides are equal and all the angles are right angles (90°).

The diagonals of a square bisect each other (cut in half) at right angles.

The diagonals of a square bisect the angles of the square.

The angles of a triangle add up to 180°.

An angle formed at the circumference of a semicircle is a right angle.

8 How much? – Money

Reminders

1. 100 pence (p) = £1
2. Pence are separated from £s by decimal points.
3. Keep the points under each other and work as with ordinary numbers.

Multiplication

We will start by multiplying pence:

EXAMPLE 1   83p x 7

$$\begin{array}{r} 83p \\ \underline{x7} \\ 581p \end{array}$$ You should already know that this result can be written as **£5.81**.

REASON. The **81** represents **tens** and **units** and the decimal point is used to separate the tens and units from the **hundreds (5)** which make whole £s. (100p = £1)

We will now try with pence and pounds:

EXAMPLE 2   £3.45 x 5

$$\begin{array}{r} £3.45 \\ \underline{x5} \\ £17.25 \end{array}$$ When you multiply the number 345 by 5 you get 1725. If we did this, it would be like multiplying **345p** by 5 and the answer 1725 would also be in pence. But only the last two figures in the sum are pence (45p) and therefore only **the last two figures in the answer are pence**. A decimal point must be put in before the last two figures to separate the pence from the pounds.

And now an example in long multiplication:

EXAMPLE 3   £23.68 x 24

Starting with left-hand figure:

$$\begin{array}{r} £23.68 \\ \underline{x24} \\ 47360 \\ \underline{9472} \\ 56832 \end{array}$$

Starting with right-hand figure:

$$\begin{array}{r} £23.68 \\ \underline{x24} \\ 9472 \\ \underline{47360} \\ 56832 \end{array}$$

The decimal point now has to be put into the answer. If we want it to be in £s, where should the point go?

Exercise 62

1. 35p
  x4
£ ___
	2. 43p
	  x6
	£ ___
		3. 52p
		  x7
		£ ___
			4. 66p
			  x3
			£ ___

5. 71p
  x3
£ ___
	6. 85p
	  x7
	£ ___
		7. 92p
		  x8
		£ ___
			8. 27p
			  x9
			£ ___

9. 32p
  x7
£ ___
	10. 46p
	  x5
	£ ___
		11. 57p
		  x4
		£ ___
			12. 63p
			  x6
			£ ___

13. £2.12 x 6	**14.** £3.14 x 5	**15.** £2.24 x 3	**16.** £3.34 x 4
17. £4.40 x 6	**18.** £4.60 x 4	**19.** £4.40 x 7	**20.** £4.70 x 4
21. £3.21 x 8	**22.** £3.31 x 9	**23.** £3.83 x 4	**24.** £4.95 x 4
25. £6.78 x 3	**26.** £6.78 x 4	**27.** £7.89 x 3	**28.** £7.89 x 4
29. £2.34 x 6	**30.** £2.34 x 7	**31.** £2.34 x 8	**32.** £2.34 x 9
33. £3.12 x 5	**34.** £3.50 x 6	**35.** £3.75 x 8	**36.** £3.82 x 7
37. £4.28 x 8	**38.** £5.69 x 7	**39.** £6.78 x 6	**40.** £7.89 x 5
41. £3.69 x 5	**42.** £4.59 x 6	**43.** £5.49 x 7	**44.** £6.39 x 8
45. £12.12 x 4	**46.** £21.34 x 6	**47.** £34.45 x 7	**48.** £43.63 x 8
49. £21.03 x 5	**50.** £32.08 x 4	**51.** £64.09 x 6	**52.** £73.07 x 3
53. £2.41 x 16	**54.** £3.16 x 14	**55.** £4.25 x 18	**56.** £6.08 x 17
57. £5.62 x 15	**58.** £7.12 x 18	**59.** £8.44 x 16	**60.** £2.73 x 19
61. £12.42 x 21	**62.** £21.24 x 23	**63.** £18.35 x 18	**64.** £24.56 x 24
65. £22.72 x 22	**66.** £16.53 x 25	**67.** £26.18 x 16	**68.** £28.22 x 15
69. £32.46 x 14	**70.** £23.61 x 21	**71.** £25.73 x 24	**72.** £31.82 x 27

Division

EXAMPLE 1   £3.84 ÷ 16

16)£3.84

16)£3.84

   0.24
16)£3.84
   3 2↓
    64
    64

Steps:
1. The number outside the bracket is already a whole number so there is no need to move decimal points.
2. Decimal point placed in answer above point in sum.
3. Divide as a long division. 16 into 3 'goes nought times' so a nought (0) goes into the answer. Carry on from there —
4. What is the answer 0.24?
5. It could be given as £0.24 or 24p

EXAMPLE 2   £68.82 ÷ 62

       £1.11
62)£68.82
    62↓
    68
    62↓
    62
    62   Answer £1.11

Exercise 63

1. £5.76 ÷ 18	**2.** £8.64 ÷ 24	**3.** £6.75 ÷ 15	**4.** £5.04 ÷ 14
5. £4.48 ÷ 16	**6.** £5.78 ÷ 17	**7.** £9.24 ÷ 22	**8.** £8.40 ÷ 15
9. £9.12 ÷ 24	**10.** £9.52 ÷ 14	**11.** £14.40 ÷ 30	**12.** £16.64 ÷ 32
13. £17.28 ÷ 36	**14.** £20.14 ÷ 38	**15.** £15.68 ÷ 14	**16.** £19.52 ÷ 16
17. £36.72 ÷ 17	**18.** £45.76 ÷ 22	**19.** £82.08 ÷ 24	**20.** £148.05 ÷ 35
21. £25.62 ÷ 42	**22.** £40.81 ÷ 53	**23.** £52.51 ÷ 59	**24.** £587.65 ÷ 73

19 How many? – Decimals

We should now be quite used to using a decimal point to separate whole pounds (£s) from the smaller parts (pence). We have also used the decimal point in measuring length, for example 18 mm = 1·8 cm; 25 cm = 2·5 dm.

The same method can be used with ordinary numbers to separate whole numbers from fractions. (Parts of whole numbers.)

The following table shows the fractions and their names:

NUMBERS	NAMES	HUNDREDS 100	TENS 10	UNITS 1	TENTHS 1/10	HUNDREDTHS 1/100	THOUSANDTHS 1/1000
(a) 4·7				4	7		
(b) 5·16				5	1	6	
(c) 23·34			2	3	3	4	
(d) 36·582			3	6	5	8	2
(e) 247·319		2	4	7	3	1	9
(f) 103·407		1	0	3	4	0	7
(g) 210·068		2	1	0	0	6	8
(h) 100·001		1	0	0	0	0	1

Note that the thick line is placed where the decimal point occurs, and in the same way separates the whole numbers from the fractions.

How we say the numbers:

(a) four point seven
(b) five point one six
(c) twenty-three point three four
(d) thirty-six point five eight two
(e) two hundred and forty-seven point three one nine
(f) one hundred and three point four nought seven
(g) two hundred and ten point nought six eight
(h) one hundred point nought nought one

Exercise 64

Write the following numbers in figures:

1. six point two
2. five point nine
3. seven point three
4. nine point one
5. eight point six
6. two point seven
7. three point nought two
8. four point nought three
9. seven point nought one
10. six point nought eight
11. thirty-two point four
12. forty-one point six
13. fifty-three point nine
14. sixty-two point three
15. seventy-eight point one
16. eighty-seven point two
17. one hundred and forty-six point seven
18. three hundred and fifty-eight point four
19. two hundred and twenty-six point one five
20. four hundred and thirty-seven point two six
21. five hundred and forty-nine point nought three
22. one hundred and seven point nought four
23. two hundred point five three one
24. four hundred point nought nought two

Multiplying decimals by 10, 100, 1000, etc

EXAMPLE   (a) 3·6 x 10 = 36
reason: 3 x 10 makes 30

More examples:
(b) 13·6 x 10   = 136
(c) 1·36 x 10   = 13·6
(d) 4·52 x 100   = 452
(e) 5·6 x 100   = 560
(f) 24·2 x 1000 = 24000

RULES   To multiply by 10, move the decimal point one place to the right
         To multiply by 100, move the decimal point two places to the right
         To multiply by 1000, move the decimal point three places to the right

NOTE   **1.** The number of noughts in 10; 100; 1000 tells you how many places to
        move the decimal point. (That is 1 or 2 or 3)
        2. Sometimes you may have to put noughts on the end of the number in order
        to be able to move the decimal point the right number of places. (See
        examples e and f)

Exercise 65

1. $5 \cdot 6 \times 10$	**2.** $7 \cdot 2 \times 10$	**3.** $8 \cdot 7 \times 10$	**4.** $9 \cdot 3 \times 10$
5. $14 \cdot 1 \times 10$	**6.** $18 \cdot 5 \times 10$	**7.** $25 \cdot 3 \times 10$	**8.** $31 \cdot 9 \times 10$
9. $2 \cdot 63 \times 10$	**10.** $5 \cdot 82 \times 10$	**11.** $8 \cdot 76 \times 10$	**12.** $9 \cdot 14 \times 10$
13. $4 \cdot 04 \times 10$	**14.** $6 \cdot 01 \times 10$	**15.** $5 \cdot 03 \times 10$	**16.** $7 \cdot 05 \times 10$
17. $4 \cdot 04 \times 100$	**18.** $6 \cdot 01 \times 100$	**19.** $5 \cdot 03 \times 100$	**20.** $7 \cdot 05 \times 100$
21. $2 \cdot 63 \times 100$	**22.** $5 \cdot 82 \times 100$	**23.** $8 \cdot 76 \times 100$	**24.** $9 \cdot 14 \times 100$
25. $14 \cdot 1 \times 100$	**26.** $18 \cdot 5 \times 100$	**27.** $25 \cdot 3 \times 100$	**28.** $31 \cdot 9 \times 100$
29. $5 \cdot 6 \times 100$	**30.** $7 \cdot 2 \times 100$	**31.** $8 \cdot 7 \times 100$	**32.** $9 \cdot 3 \times 100$
33. $1 \cdot 456 \times 1000$	**34.** $5 \cdot 731 \times 1000$	**35.** $3 \cdot 215 \times 1000$	**36.** $8 \cdot 126 \times 1000$
37. $2 \cdot 605 \times 1000$	**38.** $7 \cdot 101 \times 1000$	**39.** $9 \cdot 001 \times 1000$	**40.** $6 \cdot 005 \times 1000$
41. $6 \cdot 21 \times 1000$	**42.** $1 \cdot 71 \times 1000$	**43.** $1 \cdot 94 \times 1000$	**44.** $5 \cdot 06 \times 1000$
45. $4 \cdot 1 \times 1000$	**46.** $6 \cdot 5 \times 1000$	**47.** $7 \cdot 8 \times 1000$	**48.** $1 \cdot 2 \times 1000$
49. $0 \cdot 12 \times 100$	**50.** $0 \cdot 34 \times 10$	**51.** $0 \cdot 301 \times 100$	**52.** $0 \cdot 105 \times 1000$
53. $0 \cdot 05 \times 100$	**54.** $0 \cdot 08 \times 1000$	**55.** $0 \cdot 06 \times 100$	**56.** $0 \cdot 02 \times 1000$
57. $0 \cdot 05 \times 10$	**58.** $0 \cdot 002 \times 1000$	**59.** $0 \cdot 08 \times 100$	**60.** $0 \cdot 04 \times 10$

Dividing decimals by 10, 100, 1000

EXAMPLE   (a) $36 \div 10 = 3 \cdot 6$
reason: $30 \div 10$ makes 3

More examples:

(b) $13 \overset{\frown}{6} \div 10$ $= 1 3 \cdot 6$
(c) $1 \overset{\frown}{3} \cdot 6 \div 10$ $= 1 \cdot 3 6$
(d) $4 \overset{\frown}{5 2} \div 1 0 0$ $= 4 \cdot 5 2$
(e) $5 \overset{\frown}{6 0} \div 1 0 0$ $= 5 \cdot 6$
(f) $1 2 \overset{\frown}{3 4 5} \div 1 0 0 0$ $= 1 2 \cdot 3 4 5$
(g) $\overset{\frown}{1 5} \cdot 2 \div 1 0 0 0$ $= 0 \cdot 0 1 5 2$

RULES   To divide by 10, move the decimal point one place to the left
         To divide by 100, move the decimal point two places to the left
         To divide by 1000, move the decimal point three places to the left

1. The number of noughts in 10; 100; 1000 tells you how many places to move the decimal point. (That is 1 or 2 or 3)

  2. Sometimes you may have to put some noughts at the front of the number in order to be able to move the decimal point the right number of places. (See example g)

  3. A decimal point should not be used on its own at the front of a number, say like ·5. In such a case a nought is placed in front of the point to show that there are *no whole numbers,* 0·5. (See example g)

  4. In numbers like 4·20, the final nought can be left off to give 4·2 because the value is four point two nought or *four point two* and *not* four point twenty

Exercise 66

1. 65 ÷ 10	**2.** 72 ÷ 10	**3.** 35 ÷ 10	**4.** 84 ÷ 10
5. 16·2 ÷ 10	**6.** 21·3 ÷ 10	**7.** 33·6 ÷ 10	**8.** 41·7 ÷ 10
9. 231 ÷ 10	**10.** 536 ÷ 10	**11.** 478 ÷ 10	**12.** 693 ÷ 10
13. 231 ÷ 100	**14.** 536 ÷ 100	**15.** 478 ÷ 100	**16.** 693 ÷ 100
17. 3214 ÷ 100	**18.** 5317 ÷ 100	**19.** 6415 ÷ 100	**20.** 7328 ÷ 100
21. 5326 ÷ 1000	**22.** 4729 ÷ 1000	**23.** 7356 ÷ 1000	**24.** 8619 ÷ 1000
25. 3721 ÷ 1000	**26.** 6513 ÷ 1000	**27.** 9345 ÷ 1000	**28.** 4718 ÷ 1000
29. 356·2 ÷ 1000	**30.** 274·9 ÷ 1000	**31.** 536·7 ÷ 1000	**32.** 618·9 ÷ 1000
33. 21·3 ÷ 100	**34.** 36·5 ÷ 100	**35.** 87·4 ÷ 100	**36.** 39·6 ÷ 100
37. 42·1 ÷ 1000	**38.** 73·1 ÷ 1000	**39.** 54·6 ÷ 1000	**40.** 82·3 ÷ 1000

Addition and subtraction of decimals

EXAMPLE 1   Add the following: 12·01, 100·1, 40·92, 6·105, 0·106

```
 12·01
100·1
 40·92
  6·105
  0·106
159·241   Answer 159·241
```

EXAMPLE 2   From 120·007 subtract 103·098

```
120·007
103·098
 16·909   Answer 16·909
```

NOTE   **1.** You must keep the decimal points and figures in their correct columns.

    2. Any spaces between figures or between figures and the decimal point must be marked by noughts.

Exercise 67   The following are additions:

1. 50·13	**2.** 67·24	**3.** 8·012	**4.** 19·46
6·072	9·003	24·63	6·372
14·35	10·15	7·067	35·63
0·681	3·632	83·001	17·05

5. 38·03	**6.** 43·17	**7.** 28·42	**8.** 11·005
16·912	50·006	30·067	40·18
10·007	0·192	1·232	0·952
72·106	23·07	78·009	36·107

9. 72·06	**10.** 63·87	**11.** 92·04	**12.** 24·72
19·472	28·406	31·007	63·005
0·809	7·005	0·708	31·06
26·68	43·68	22·56	10·08

The following are subtractions:

13. 23·763	**14.** 37·312	**15.** 41·135	**16.** 52·613
16·105	25·534	37·213	41·935

17. 64·321	**18.** 75·417	**19.** 83·125	**20.** 91·595
56·219	61·329	74·317	83·618

21. 101·79	**22.** 215·06	**23.** 306·83	**24.** 405·09
92·05	109·73	207·09	391·06

Multiplication of decimals

EXAMPLE 1   Multiply 8·26 by 34

Starting with the left-hand figure:

```
  826
  x34
24780
 3304
28084
```

Starting with the right-hand figure:

```
  826
  x34
 3304
24780
28084
```

280·84   Answer 280·84

Steps: (1) Multiply as if using whole numbers.

(2) Count the number of figures after the decimal points in the sum: 8·26. There are two (2) such figures.

(3) Count two places from the right in the answer and place the decimal point.

EXAMPLE 2   Multiply 64·37 by 28·5

Starting with the left-hand figure:

```
   6437
    285
1287400
 514960
  32185
1834545
```

Starting with the right-hand figure:

```
   6437
    285
  32185
 514960
1287400
1834545
```

1 8 3 4 · 5 4 5   Answer 1834·545

Steps: As for Example 1, but this time there are three (3) figures after the decimal points in the sum: 64·37 x 28·⑤. Therefore count three places from the right in the answer and place the decimal point.

Exercise 68

1. 2·9 x 1·2	**2.** 3·4 x 1·3	**3.** 4·2 x 1·5	**4.** 3·7 x 1·4
5. 2·7 x 1·6	**6.** 3·2 x 1·7	**7.** 4·5 x 1·3	**8.** 3·6 x 1·5
9. 5·9 x 1·2	**10.** 5·1 x 1·8	**11.** 6·2 x 1·4	**12.** 6·7 x 1·6
13. 17·3 x 1·3	**14.** 23·2 x 1·5	**15.** 1·73 x 1·3	**16.** 2·32 x 1·5
17. 0·173 x 1·5	**18.** 0·173 x 0·15	**19.** 0·232 x 1·3	**20.** 0·232 x 0·13
21. 0·392 x 12	**22.** 39·2 x 1·2	**23.** 0·006 x 6	**24.** 0·6 x 0·06
25. 12·4 x 14	**26.** 1·24 x 1·4	**27.** 124 x 0·14	**28.** 12·4 x 0·14
29. 28·4 x 2·4	**30.** 4·62 x 3·6	**31.** 93 x 0·81	**32.** 5·23 x 10·2
33. 62·4 x 2·34	**34.** 234 x 0·143	**35.** 4·17 x 3·56	**36.** 0·32 x 2·14

Division of decimals

EXAMPLE 1   Divide 5·832 by 1·8

5·832 ÷ 1·8

1·8)5·832

18)58·32

Steps:

1. Turn the number *outside* the bracket into a **whole number** by moving the decimal point to the extreme right, it moves one place. (1·8 becomes 18)

2. Move the decimal point the **same number of places** (one) for the number *inside* the bracket. (5·832 becomes 58·32)

3. The answer will appear above the line, place a decimal point in the answer above the point in the sum.

4. Now divide as a long division sum.

$$18\overline{)58\cdot32}$$

$$\begin{array}{r} 3\cdot24 \\ 18\overline{)58.32} \\ \underline{54}\downarrow \\ 43 \\ \underline{36}\downarrow \\ 72 \\ \underline{72} \end{array}$$ Answer 3·24

EXAMPLE 2   Divide 49·504 by 1·36

$$49\cdot504 \div 1\cdot36$$

$$1\cdot36\overline{)49\cdot504}$$

$$136\overline{)4950\cdot4}$$

$$136\overline{)4950\cdot4}$$

Steps:
1. Turn 1·36 into 136; the decimal point moves two places.
2. Move the decimal point two places in 49·504; it becomes 4950·4.
3. Put the decimal point in the answer above the point in the sum.
4. Divide as a long division.

$$\begin{array}{r} 36\cdot4 \\ 136\overline{)4950\cdot4} \\ \underline{408}\downarrow \\ 870 \\ \underline{816}\downarrow \\ 544 \\ \underline{544} \end{array}$$ Answer 36·4

Exercise 69

1. 1·92 ÷ 1·2	**2.** 1·68 ÷ 1·4	**3.** 2·4 ÷ 1·6	**4.** 2·7 ÷ 1·5
5. 2·24 ÷ 1·4	**6.** 2·34 ÷ 1·8	**7.** 1·8 ÷ 1·2	**8.** 2·21 ÷ 1·7
9. 3·36 ÷ 2·4	**10.** 3·25 ÷ 1·3	**11.** 4·16 ÷ 2·6	**12.** 3·36 ÷ 1·2
13. 4·8 ÷ 3·2	**14.** 6·48 ÷ 1·8	**15.** 5·32 ÷ 3·8	**16.** 6·72 ÷ 1·6
17. 5·06 ÷ 2·2	**18.** 6 ÷ 2·4	**19.** 5·88 ÷ 2·8	**20.** 7·02 ÷ 2·6
21. 14·64 ÷ 1·2	**22.** 20·44 ÷ 1·4	**23.** 19·56 ÷ 1·2	**24.** 25·76 ÷ 1·4
25. 17·92 ÷ 0·8	**26.** 14·52 ÷ 0·6	**27.** 12·2 ÷ 0·4	**28.** 6·52 ÷ 0·2
29. 0·56 ÷ 0·5	**30.** 0·848 ÷ 0·4	**31.** 1·884 ÷ 0·6	**32.** 1·272 ÷ 0·3
33. 0·288 ÷ 0·12	**34.** 0·364 ÷ 0·14	**35.** 0·308 ÷ 0·11	**36.** 0·576 ÷ 0·18
37. 0·425 ÷ 2·5	**38.** 0·324 ÷ 0·12	**39.** 0·476 ÷ 3·4	**40.** 0·54 ÷ 0·15
41. 2·296 ÷ 16·4	**42.** 2·912 ÷ 0·16	**43.** 3·996 ÷ 22·2	**44.** 5·83 ÷ 0·22
45. 7·268 ÷ 3·16	**46.** 13·696 ÷ 3·2	**47.** 23·848 ÷ 5·42	**48.** 350·56 ÷ 5·6

0 How much? – Weight

The unit of weight is the **gram** and the abbreviation for gram is **g.**

When measuring weight we use a decimal point to separate whole ones from fractions just as we do with money and measuring length.

There are other units of measuring weight, some smaller than the gram, others larger. We will do a little work using only the gram for the present time.

XAMPLE 1   Add the following weights:

g
102·68
256·7
 32·83
 41·004
433·214   Answer 433·214 g

XAMPLE 2   Subtract 385·671 g from 514·52 g

g
514·52
385·671
128·849   Answer 128·849 g

XAMPLE 3   Multiply 38·42 g by 4·5

3842 g
  x45
153680   Where does the decimal point go?
 19210
172890   Answer ? g

XAMPLE 4   Divide 220·32 g by 3·4

220 ÷ 3·4 becomes 2203·2 ÷ 34

```
        64·8
34) 2203·2
    204
    163
    136
    272
    272   Answer 64·8 g
```

Exercise 70   The following are additions:

g	g	g	g
1. 12·68	**2.** 204·3	**3.** 120·64	**4.** 216·08
234·203	89·607	95·407	58·406
96·15	123·98	342·63	415·19
152·248	70·092	63·904	92·007
____ g	____ g	____ g	____ g

g	g	g	g
5. 67·501	**6.** 36·005	**7.** 8·072	**8.** 61·43
0·988	100·98	89·009	205·097
235·47	97·067	416·91	86·235
9·007	232·19	0·989	160·05
____ g	____ g	____ g	____ g

g	g	g	g
9. 231·07	**10.** 21·074	**11.** 9·507	**12.** 29·008
76·102	8·316	160·43	0·982
504·09	302·09	43·704	317·07
10·009	67·125	208·07	76·001
____ g	____ g	____ g	____ g

The following are subtractions:

g	g	g	g
13. 87·312	**14.** 70·115	**15.** 92·056	**16.** 64·203
25·041	43·082	71·347	37·106
____ g	____ g	____ g	____ g

g	g	g	g
17. 36·05	**18.** 43·152	**19.** 5·072	**20.** 75·106
25·161	16·007	0·783	54·009
____ g	____ g	____ g	____ g

The following are multiplications:

21. 3·52 g x 14 **22.** 41·2 g x 12 **23.** 0·68 g x 22 **24.** 5·18 g x 16
25. 7·06 g x 18 **26.** 206 g x 2·4 **27.** 325 g x 0·42 **28.** 416 g x 5·6
29. 524 g x 0·52 **30.** 38·7 g x 3·5 **31.** 60·8 g x 4·6 **32.** 5·06 g x 6·2

The following are divisions:

33. 293·4 g ÷ 18 **34.** 387·2 g ÷ 16 **35.** 439·6 g ÷ 14 **36.** 11·704 ÷ 2·2
37. 97·68 g ÷ 12 **38.** 7·242 g ÷ 1·7 **39.** 173·46 g ÷ 4·2 **40.** 32·344 g ÷ 5·2
41. 19·248 g ÷ 0·24 **42.** 0·3392 g ÷ 1·6 **43.** 1·6448 g ÷ 0·32 **44.** 0·1008 g ÷ 0·36